Laboratory Techniques in Membrane Biophysics

An Introductory Course

By

W. McD. Armstrong · K. Baumann · P. C. Caldwell
T. W. Clarkson · J. Dudel · B. Frankenhaeuser · E. Frömter · G. Gardos
B. Z. Ginzburg · J. F. Hoffman · A. Kepes · R. D. Keynes
A. Kleinzeller · P. G. Kostyuk · A. Kotyk · A. A. Lev · B. Lindemann
P. Mueller · H. Muth · W. Nonner · E. Oberhausen · H. Passow
A. Rothstein · D. O. Rudin · R. Stämpfli · T. Teorell · K. J. Ullrich
N. A. Walker · W. Wilbrandt

Edited by H. Passow and R. Stämpfli

With 66 Figures

D1444208

Springer-Verlag Berlin · Heidelberg · New York 1969

Title No. 1553

Preface

The present manual contains a collection of laboratory instructions used during an international training course on membrane biophysics which was held at Homburg in the fall of 1966. The selection of the topics dealt with in the various chapters depended on the scientific interest of the available teachers and on the availability of the necessary equipment in our laboratories. Thus, the material included in this volume does not add up to a systematic course in membrane biophysics. Instead it represents a more fortuitous collection of laboratory problems. In addition, some authors place more emphasis on teaching the more technical aspects of a method whereas others are primarily concerned with the demonstration of a significant biological phenomenon. Nevertheless, in spite of such differences of emphasis and a somewhat haphazard choice of a few methods and phenomena among many others of similar importance, it was felt that the publication of the material is desirable. Since no other laboratory manual exists so far, the present laboratory problems which were tested in actual practice may serve as a useful basis for the shaping of further training courses or for laboratory courses for graduate students in biophysics and physiology.

Our thanks are due to the authors and the publisher who were patient and kind enough to cooperate with the editors during the long period between the end of the course and the appearance of the book. We are particularly grateful to a number of authors who were willing to expand their material beyond the limits of their original manuscripts.

Homburg, February 1969   H. Passow · R. Stämpfli

Contents

List of Authors

ARMSTRONG, W. McD.: Department of Physiology, Indiana University, Indianapolis (USA)

BAUMANN, K.: Physiologisches Institut der Freien Universität, Berlin (Germany)

CALDWELL, P. C.: Department of Zoology, University of Bristol (Great Britain)

CLARKSON, T. W.: Department of Radiation Biology and Biophysics, University of Rochester, Rochester (USA)

DUDEL, J.: II. Physiologisches Institut der Universität Heidelberg, Heidelberg (Germany)

FRANKENHAEUSER, B.: Nobel Institute for Neurophysiology, Karolinska Institutet, Stockholm (Sweden)

FRÖMTER, E.: Physiologisches Institut der Freien Universität, Berlin (Germany)

GARDOS, G.: Department of Cell Metabolism, Research Institute of the National Blood Center, Budapest (Hungary)

GINZBURG, B. Z.: Department of Botany, Hebrew University, Jerusalem (Israel)

HOFFMAN, J. F.: Department of Physiology, Yale University, New Haven (USA)

KEPES, A.: Collège de France, Laboratoire de Biologie Moléculaire, Paris (France)

KEYNES, R. D.: Institute of Animal Physiology, Agricultural Research Council, Babraham, Cambridge (Great Britain)

KLEINZELLER, A.: Department of Physiology, University of Pennsylvania, Philadelphia (USA)

KOSTYUK, P. G.: Institute of Physiology, Kiev (USSR)

KOTYK, A.: Institute of Microbiology, Czechoslovak Academy of Sciences, Prague (Czechoslovakia)

LEV, A. A.: Institute of Cytology, Academy of Sciences of USSR, Leningrad (USSR)

LINDEMANN, B.: II. Physiologisches Institut, Universität des Saarlandes, Homburg a. d. Saar (Germany)

MUELLER, P.: Eastern Pennsylvania Psychiatric Institute, Philadelphia (USA)

MUTH, H.: Institut für Biophysik, Universität des Saarlandes, Homburg a. d. Saar (Germany)

NONNER, W.: I. Physiologisches Institut, Universität des Saarlandes, Homburg a. d. Saar (Germany)

OBERHAUSEN, E.: Institut für Biophysik, Universität des Saarlandes, Homburg a. d. Saar (Germany)

PASSOW, H.: II. Physiologisches Institut, Universität des Saarlandes, Homburg a. d. Saar (Germany)

ROTHSTEIN, A.: Department of Radiation Biology and Biophysics, University of Rochester, Rochester (USA)

RUDIN, D. O.: Department of Basic Research, Eastern Pennsylvania Psychiatric Institute, Philadelphia, Pennsylvania (USA)

STÄMPFLI, R.: I. Physiologisches Institut, Universität des Saarlandes, Homburg a. d. Saar (Germany)

TEORELL, T.: Department of Physiology, University of Uppsala, Uppsala (Sweden)

ULLRICH, K. J.: Physiologisches Institut der Freien Universität, Berlin (Germany)

WALKER, N. A.: School of Biological Sciences, University of Sydney, Sydney (Australia)

WILBRANDT, W.: Pharmakologisches Institut, Universität Bern, Bern (Switzerland)

Measurement of Hydraulic Conductivity and Reflexion Coefficients of a Plant Cell Membrane by means of Transcellular Osmosis

By B. Z. Ginzburg and N. A. Walker

I. Introduction

The aim of this experiment is to measure the hydraulic conductivity (L_P) of the "membrane" of the *Nitella* internodal cell, and its reflexion coefficient (σ) for a rapidly penetrating solute. The coefficients L_P and σ appear in the phenomenological equations for the movement of water and solute across a homogeneous membrane separating solutions of the same nonelectrolyte. It is possible (KEDEM and KATCHALSKY, 1958) to write the dissipation function as:

$$\Phi = J_V \cdot \Delta P + J_D \cdot \Delta \Pi \tag{1}$$

where

J_V is the flow of volume across the membrane,

J_D is the relative velocity of solute to solvent in the membrane,

ΔP is the hydrostatic pressure difference across the membrane,

$\Delta \Pi$ is the osmotic pressure difference across the membrane.

Making the assumption that flows are linear functions of forces,

$$J_V = L_P \cdot \Delta P + L_{PD} \cdot \Delta \Pi \tag{2}$$

$$J_D = L_{DP} \cdot \Delta P + L_D \cdot \Delta \Pi . \tag{3}$$

And Eq. (2) can be rewritten

$$J_V = L_P \cdot \Delta P - \sigma \cdot L_P \cdot \Delta \Pi \tag{4}$$

where σ is defined as $(-L_{PD}/L_P)$.

At $\Delta \Pi = 0, L_P$ clearly measures the relation between hydrostatic pressure difference and volume flow. It is a kinetic parameter (DENBIGH 1951) and can only be obtained by a measurement of flow and driving force. At $J_V = 0, \sigma$ measures the "relative effectiveness" of hydrostatic and osmotic pressure differences: it is a ratio of two similar kinetic parameters — i.e. a thermodynamic parameter. It can be measured by a nul method (at $J_V = 0$). Its meaning can be grasped from this method

1 Membrane Biophysics

of measurement, and from the relation $\sigma = 1 - \dfrac{v_s}{v_w}$, where v_s is the velocity of the solute in the membrane and v_w the velocity of water. Normally then values of σ will range from 1, for impermeant solutes $(v_s = 0)$ to 0, for solute permeating as readily as water $(v_s = v_w)$; though negative values of σ are possible (Katchalsky and Curran, 1965).

In this experiment L_P is measured by volume flow and σ by the nul method.

Transcellular osmosis was first observed by Osterhout, and it has been fully treated by Kamiya and Tazawa (1956) and by Dainty and Hope (1959). The treatment below is somewhat simplified.

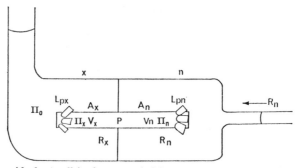

Fig. 1. A turgid plant cell is shown, sealed between two compartments x, n. Compartment x is open to the atmosphere, while n is also open via the capillary. Changes in volume of the contents of compartment n may be measured by the movement of the meniscus in the capillary. Meaning of symbols: Π_o osmotic pressure of solution in compartment x. $\Pi_{x,n}$ osmotic pressure of solution in cell, at end in compartment x, n. P hydrostatic pressure of cell contents. $A_{x,n}$ area of cell surface in compartment x, n. $V_{x,n}$ volume of cell in compartment x, n. $R_{x,n}$ rate of volume flow out of/into cell in compartment x, n. R_m net rate of change of volume in compartment n

A turgid plant cell (Fig. 1) is sealed between two compartments (here labelled x and n) so that the areas and volumes of the cell exposed in the two compartments are A_x, V_x; A_n, V_n respectively. The hydrostatic pressure in the cell interior is P, while the osmotic pressures of the solutes in the cell interior are Π_x, Π_n respectively. The exposed cell surfaces have values of L_P denoted by L_{Px} and L_{Pn} respectively.

We imagine an experiment in which initially there is water on both sides, and no net water movement, $\Pi_o = 0$, $\Pi_x = \Pi_n$: at zero time the water on side x is replaced by sucrose solution of osmotic pressure Π_o, and the rate of volume change R_m in compartment n is measured. It will be assumed that for sucrose $\sigma = 1$.

At end x of the cell water will move from cell interior to solution at a rate:

$$R_x = L_{Px} A_x (\Pi_o - \Pi_x + P) \tag{5}$$

and since the exit of water will reduce P below its initial value there will be a flow of water into end n of the cell from compartment n:

$$R_n = L_{Pn} A_n (\Pi_n - P) . \qquad (6)$$

Water will move along the cell interior from n to x, sweeping internal solutes towards end x, and for this reason we write the osmotic pressures of internal solutes as Π_n and Π_x.

If $R_n \neq R_x$ the cell will change in volume at a rate $(R_n - R_x)$. Now in this experiment one measures the rate of net volume change (R_m) in compartment n, and this rate is composed of R_n and the change in volume of that part of the cell in compartment n, viz:

$$R_m = R_n - (R_n - R_x) \cdot \left(\frac{V_n}{V_n + V_x} \right) \qquad (7)$$

and if the cell is cylindrical $\dfrac{V_n}{V_n + V_x} = \dfrac{A_n}{A_n + A_x}$, so

$$R_m = R_n - (R_n - R_x) \cdot \left(\frac{A_n}{A_n + A_x} \right) . \qquad (8)$$

Inserting values from Eqs. (5) and (6):

$$R_m = \frac{A_x A_n}{A_x + A_n} L_{Px} \Pi_0 - \frac{A_x A_n}{A_x + A_n} [L_{Px}(\Pi_x - P) - L_{Pn}(\Pi_n - P)] . \qquad (9)$$

Calculation from actual rates of flow and cell dimensions shows that during the first min of transcellular osmosis the volume of water flowing through the cell is about 2% of its volume, so that if we restrict measurements to this period we can assume $\Pi_i \doteqdot \Pi_n$, and if $\Pi_i = \Pi_x = \Pi_n$ then

$$R_m \doteqdot \frac{A_x A_n}{A_x + A_n} L_{Px} \Pi_0 - \frac{A_x A_n}{A_x + A_n} [(L_{Px} - L_{Pn}) (\Pi_i - P)] . \qquad (10)$$

Further simplification, which is not strictly justified, may be introduced for the purposes of this one-day experiment: we assume $L_{Px} = L_{Pn}$ and obtain:

$$R_m \doteqdot \frac{A_x A_n}{A_x + A_n} L_P \Pi_0 . \qquad (11)$$

This will give the rate at any moment after zero time; whether due to cell shrinkage or to water flow from one compartment to the other, until Π_x is significantly different from Π_n. From this equation, having measured R_m, A_x, A_n and Π_0, we can obtain L_P.

If the solute used (say e) has a value of σ not equal to 1, a similar argument will show that

$$R_m \doteqdot \frac{A_x A_n}{A_x + A_n} L_P \cdot \sigma_e \cdot RTC_e \qquad (12)$$

where C_e is the concentration of the solute at side x. If R_m, A_x, A_n, C_e and L_P are known, σ_e may be calculated.

1*

Alternatively if we have sucrose at concentration C_s on side n and find a concentration of solute C_e at side x for which $R_m = 0$ initially, we can show (again assuming $L_{Px} = L_{Pn} = L_P$ and $\sigma_s = 1$)

$$R_m = 0 = R_n - (R_x - R_n) \cdot \left(\frac{V_n}{V_n + V_x}\right) \qquad \text{[Eq. (7)]}$$

$$= \frac{A_x A_n}{A_x + A_n} RTL_P(\sigma_e C_e - C_s)$$

whence $\sigma_e = C_s/C_e$.   (13)

II. Apparatus and Solutions

1 Transcellular osmosis apparatus (Fig. 2).

1 Large stirred waterbath to accommodate apparatus and twelve 100 ml bottles of solution.

12 100 ml bottles.

1 Stopwatch.

1 Mirror scale (mm).

1 Microscope with $10 \times$ objective and $10 \times$ micrometer exepiece.

3 20 ml syringes with flexible plastic tubing attached.

"Kleenex".

Solutions of sucrose: 0.1, 0.2, 0.3, 0.4, 0.6, 0.8 molal

and $\begin{cases} \text{isopropanol: 0.1, 0.2, 0.4, 0.6, 0.8, 1.0 molal} \\ \text{or ethanol: same concentrations.} \end{cases}$

"Vaseline".

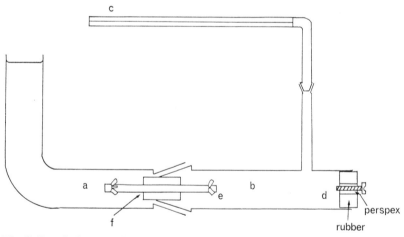

Fig. 2. Practical transcellular osmosis apparatus. a, b two halves of apparatus containing solutions. c capillary, of 0.2 mm i.d. if mirror scale is used or 1.0 mm if travelling microscope is used. d rubber bung with nylon screw insert for adjusting zero position of capillary. e internodal cell of *Nitella* or *Chara*. f split bung of "perspex" or rubber, with groove just larger than cell

Plants of *Nitella translucens, Nitella flexilis, Chara corallina,* or some similar large uncorticated species.

III. Execution

a) Preparation of Apparatus. Clean the joints and coat them with vaseline. Dry the rubber or perspex split bung and adequately coat with vaseline. Clean the capillary throroughly, e.g. with detergent solution followed by distilled water.

b) Cell. Select a healthy, clean internode, perhaps 8 — 10 cm long, and cut it free of neighbouring internodes with fine scissors. Cells may be handled gently with the fingers. The selected cell is examined under the microscope and its diameter is measured at say four points along the cell; when cytoplasmic streaming is observed it can be assumed that the cell is healthy, and the experiment can proceed.

The cell is blotted gently but thoroughly with Kleenex, and vaseline is applied to its centre with dry vaselined fingers, making sure the vaseline sticks to the cell right round its circumference. This is most important, since the experiment will fail if the vaseline is not made to adhere to the dry cell wall. Immediately, dip the ends of the cell into distilled water — the cell will collapse if kept dry in air for more than a few seconds. The cell is then placed between the vaselined halves of the split bung, and the whole sealed into part *b* of the apparatus (Fig. 2). The cell has been positioned so that roughly equal areas protrude from each end of the bung.

The apparatus is assembled with similar solutions (for the first experiment, distilled water) in compartments *a* and *b*, and all large air bubbles are removed from compartment *b*. The apparatus (and the bottles of solution) are set in the water bath, and thermal equilibrium secured.

With a good seal around the cell, the movement of the meniscus due to thermometer effect should be rapid when the apparatus is first put into the bath, and should eventually be less than 1 mm/min. Continued rapid movement of the meniscus towards the end of the capillary indicates a leak round the cell (and for this reason a meniscus is preferred to a bubble in the capillary).

c) The Experiment.
1. Measurement of L_P. When thermal equilibrium is reached with distilled water at *a* and *b*, the water in *a* is replaced with 0.1 molal sucrose, and the rate of movement of the meniscus is measured for the first 60 — 100 sec.

The sucrose is replaced by distilled water and the cell is left for 5 — 15 min until the meniscus has again ceased to move. Similar flow measurements are made for each sucrose concentration.

If there is sufficient time, the whole should be repeated with 0.2 molal sucrose at b and $0.3 - 0.8$ molal at a, using 0.2 molal as the "resting" solution on side a.

2. **Measurement of σ.** Beginning with 0.1 molal sucrose at a and b, the solution at a is removed and after a rinse with distilled water is replaced with isopropanol or ethanol solution, and the initial movement of the meniscus noted (over say $0 - 20$ sec). By interpolation a concentration is predicted for which the initial movement would be zero.

Alternatively the apparatus is set up with distilled water on both sides, and initial flow rates are found for the series of isopropanol concentrations, exactly as in part (1). It is essential to have already measured L_P for the particular cell used.

At the end of the experiment the lengths of cell at a and b should be measured with a mm scale; and the cell is observed again under the microscope for protoplasmic streaming.

The capillary is calibrated by sucking a pellet of mercury into its bore, measuring the length, and then weighing the mercury.

d) **Evaluation of Data.** It is simple to calculate L_P from the relation derived from Eq. (11),

$$L_P = \frac{R_m}{RTC_a - RTC_b}\left(\frac{A_a + A_b}{A_a A_b}\right)$$

where the subscripts refer to ends a and b. Values of L_P so found are tabulated against C_a (and C_b) so that any variation of L_P with C can be seen.

In the flow method for σ, using Eqs. (11) and (12) $\sigma_e = \dfrac{R_e}{R_s}$ where R_e and R_s are rates of change of volume at the same concentrations of solute and sucrose.

In the nul method, Eq. (13) gives $\sigma_e = \dfrac{C_s}{C_e}$ where C_s and C_e are concentrations of sucrose and solute which give zero volume change.

In all calculation the osmotic coefficients of the solutes should be looked up and applied if necessary.

IV. Comments

The original papers should be read for a full discussion of the theoretical and practical difficulties of this method, since the following comments are brief.

a) **Measurement of L_P.** We have assumed here that L_P is the same for entry and for exit of water. If however this is checked experimentally, it is found not to be true. Such an experiment can be done by setting up a cell with unequal areas A_a and A_b exposed, and performing two flow measurements:

I. with sucrose at a and distilled water at b,
II. with distilled water at a and sucrose at b.

If $L_{Pn} \neq L_{Px}$ we can show

$$R_m^I = \frac{L_{Pn} L_{Px} A_a A_b}{L_{Pn} A_b + L_{Px} A_a} \Pi_0$$

$$R_m^{II} = \frac{L_{Pn} L_{Px} A_a A_b}{L_{Pn} A_a + L_{Px} A_b} \Pi_0$$

and L_{Pn}, L_{Px} can be calculated. Experimentally one finds that $L_{Pn} > L_{Px}$. A partial explanation of this result may be the effect of unstirred layers. An unstirred layer exists outside the cell membrane in the sucrose solution, and in this layer the steady flow of water will sweep sucrose away from the membrane until there is established a concentration gradient for sucrose which will result in an equal rate of diffusion towards the membrane. In this steady state the concentration of sucrose at the membrane will be less than that in the bulk solution, and the osmotic driving force on water flow will be lower than that calculated from the bulk concentration. A calculation for a steady state unstirred layer by DAINTY (1963) shows that this would account for part of the observed inequality between L_{Px} and L_{Pn}; but it seems unlikely that the full steady state unstirred layer effect develops during the first 60 sec of the experiment.

A further partial explanation may be in the observed decrease of L_P with external solute concentration, attributed by DAINTY and GINZBURG to the reduction in hydration of the membrane.

b) Measurement of σ. One of the aims in measuring σ was to compare the experimental value with that calculated for a simple pore model. If solute and solvent cross the membrane in pores and interact by mutual friction, it can be shown that

$$\sigma_s = 1 - \frac{\omega_s \bar{V}_s}{L_P} - \frac{K_s^c \cdot f_s^w}{f_s^w + f_s^c} \tag{14}$$

where
ω_s is the permeability coefficient for the solute.
$\bar{V}_s$ is its partial molar volume.
L_P is the membrane hydraulic conductivity.
K_s^c is the distribution coefficient of the solute within the membrane.
f_s^w is the frictional coefficient water-solute.
f_s^c is the frictional coefficient solute-membrane.

Clearly if there is no pore interaction the third term in this equation becomes zero.

Measurements do show that the third term differs significantly from zero. However the conclusion that this demonstrates the existence of membrane pores is undermined when we consider that in the *Nitella* experiment we are measuring σ for a complex array of cell wall and two cell membranes. The meaning of L_P and σ for such a series array is by no means clear, and they can only at the moment be regarded as lumped parameters of a complex membrane system. The fact that we are dealing with a double membrane system may also be a partial explanation of the observed inequality of L_{Pn} and L_{Px}.

V. References

DAINTY, J.: Protoplasma (Wien) **57**, 220 (1963).
—, and B. Z. GINZBURG: (1) Biochim. biophys. Acta (Amst.) **79**, 102 (1964).
— — (2) Biochim. biophys. Acta. (Amst.) **79**, 112 (1964).
— — (3) Biochim. biophys. Acta. (Amst.) **79**, 122 (1964).
— — (4) Biochim. biophys. Acta. (Amst.) **79**, 129 (1964).
—, and A. B. HOPE: Aust. J. biol. Sci. **12**, 136 (1959).
DENBIGH, K. G.: The thermodynamics of the steady state. Methuen 1951. (Reprinted 1958).
KAMIYA, N., and M. TAZAWA: Protoplasma (Wien) **46**, 394 (1956).
KATCHALSKY, A., and P. F. CURRAN: Nonequilibrium thermodynamics in biophysics. Harvard 1965.
KEDEM, O., and A. KATCHALSKY: Biochim. biophys. Acta. (Amst.) **27**, 229 (1958).

Flux Measurements in Erythrocytes

By G. Gardos, J. F. Hoffman, and H. Passow

I. Introduction

a) Technical Problem. Measurement of unidirectional fluxes at equilibrium and evaluation of data in terms of two compartment kinetics.

b) Principle of Method. Washed erythrocytes are equilibrated with isotonic solutions containing a suitably chosen concentration of the test substance whose flux is to be determined. After diffusion equilibrium is established, a small amount of labelled test substance is added and the time course of disappearance of radioactivity from the supernatant is followed. The rate constant of the penetration process is determined by fitting the curve relating radioactivity in the medium to time by a single exponential generated by an analog computer. The ion flux is calculated from the rate constant and the independently determined amounts of test substance in medium and cell water.

c) Special Application. Demonstration of the effects of pH, ionic strength, temperature, and chemical nature of penetrating anion species on rate of anion transfer across the membrane of human erythrocytes. For comparison with anion permeability, the independence of pH of a nonelectrolyte, erythritol, will also be shown.

II. Equipment and Solutions

Liquid scintillation counter;
Two thermostatically controlled water baths;
Laboratory centrifuge (1600 g);
Micro-hematocrit centrifuge;
pH-meter;
Analog computer;
Stop watch;
Transparent paper, millimetre graph paper;
Thermostatically controlled oven;
Pipettes: 0.25 ml, 0.50 ml, 1.00 ml, 2.00 ml, 5.00 ml, 10.00 ml, 15.00 ml;
Analytical balance and automatic balance;

Erlenmeyer flasks 100 ml;
Weighing boats;
Micro-hematocrit tubes and micro-burettes;
Volumetric flasks;
Water jet pump;
Isotope counting vials for radioactivity measurements;
Bulb pipettor;
Centrifuge rack;
Beaker glasses and centrifuge tubes.

Solutions. For scintillation counting: Liquid scintillator (400 ml ethanol, 600 ml toluene, 2.0 g PPO, 0.025 g POPOP).

For chloride determination: 100 ml 10 mM/l Hg $(NO_3)_2$; 21% trichloroacetic acid (TCA); 50 ml 1% diphenyl carbazone in 96% ethanol; 100 ml M/5 phosphate buffer pH 7.00; 100 ml 20 mM/l NaCl; 100 ml 0.1 N citrate buffer pH 1.5 − 2;

Test substances:

Succinic acid	133 mM/l;	NaOH	1 M/l;
Na_2SO_4	133 mM/l;	HCl	1 M/l;
erythritol	332 mM/l;		
sucrose	332 mM/l;		
NaCl	166 mM/l.		

III. Execution

Freshly drawn citrated blood (20 ml 3.8% Na-citrate per 100 ml blood) is centrifuged in 50 ml tubes at 1600 g for 15 min. The supernatant including the buffy coat is removed with a water jet pump; the cells are washed by filling the centrifuge tubes with isotonic NaCl solution, thoroughly stirring and centrifuging again. After 3 washes, 10 portions of erythrocyte sediment, 15 g each, are weighed out in 100 ml Erlenmeyer flasks which are labelled 1 − 10. 15 ml each of the following solutions are added to these flasks, such that solution 1 is placed in flask 1, solution 2 in flask 2, and so on:

No.	Na_2SO_4 mM/l	Succinate mM/l	Erythritol 1 mM/l	NaCl mM/l	Sucrose mM/l	pH	Temp. °C
1	10	—	—	83	136	7.9	37
2	10	—	—	83	136	7.0	37
3	10	—	—	83	136	7.4	37
4	10	—	—	83	136	7.4	27
5	10	—	—	55	192	7.4	37
6	10	—	—	83	136	7.4	37
7	—	10	—	83	136	7.9	37
8	—	10	—	83	136	7.0	37
9	—	—	10	166	0	7.9	37
10	—	—	10	166	0	7.0	37

pH is adjusted by addition of HCl or NaOH. The pH value in the table refers to the pH in the final suspension.

After thorough mixing, all suspensions are incubated at 37 °C for 2 h (except suspension 4, which is incubated at 27 °C) in order to establish diffusion equilibrium. Three 1 ml samples are withdrawn from each Erlenmeyer flask for pH, water content, and hematocrit determinations.

pH determination of the supernatant is done with a Radiometer Instrument. For standardization, a phosphate buffer pH 7.00 is supplied. For wet and dry weight determination, 1 ml of suspension is pipetted into a weighing boat of known weight, and the weight is determined as rapidly as possible (to avoid evaporation) on an automatic balance. The water is then evaporated by heating to 110 °C for about 5 h (i.e. until after the experimental data are fitted on the analog computer). Cell volume is determined with micro-hematocrit tubes. The tubes are filled by dipping them with forceps into a small sample of suspension. Their lower ends are closed with sealing wax. The tubes are centrifuged on a micro-hematocrit centrifuge at full speed for 10 min.

Subsequently 0.3 ml aliquots of radioactive $Na_2{}^{35}SO_4$ (flasks 1 − 6), succinate ^{14}C (flasks 7, 8), or erythritol ^{14}C (flasks 9, 10) are added and the suspensions very carefully shaken (2 min) in order to insure that the radioactivity is completely removed from the walls of the flasks. The instant of addition of radioactivity is the zero time of the experiment. The first 3.0 ml sample is withdrawn as soon as possible and the following ones after 15, 30, 60, 90, 120, 180 min respectively. After centrifugation at 1600 g for 6 min, exactly 1 ml of supernatant is sucked into a pipette by using a bulb pipettor and transferred into a centrifuge tube containing 0.5 ml 21% TCA. The TCA precipitate is removed by centrifugation (5 min at 1600 g) and 0.25 ml of the clear supernatant is added to 10 ml liquid scintillator. The vials containing the mixture are kept in the refrigerator until the last sample is ready for transfer to the cold box of the Tricarb Scintillation Counter. 30 min after inserting the vials counting is started (panel settings: Mode selector: automatic. Present count: background. Present time: 0.01 min 1 × 10². Window: A, B, potentiometers at 50 and 1000, C, D at 50 and 250, E, F at 250 and 7000 respectively. Gain: Background subtraction: adjust to predetermined background in counts/min. Sample changer: operate). The print out presents the results as follows: the sample number is indicated in the first row, the counting time in the second and in the following ones the counts/min as measured in each of the three channels.

While the samples are being counted Cl-determinations may be performed. 0.5 ml of the deproteinized supernatant of the first and last sample are transferred to a 100 ml Erlenmeyer flask and after addition of 0.5 ml citrate buffer, three drops of diphenyl-carbazone, and some distilled water, the chloride content is determined by titration with Hg $(NO_3)_2$ (for standardization 1.0 ml 20 mM/l NaCl is titrated first). The results of radioactivity measurements are plotted on graph paper whereby 1 cm should represent 1000 counts per min on the ordinate and 4 cm 90 min on the abscissa. The coordinates as well as the measured points are reproduced on transparent paper (size 10 × 10 cm) and the latter attached to the screen of an oscilloscope connected to the computer. An e-function is generated on the screen by operating the computer in the repetitive mode and after the standardization procedure outlined below the generated curve is fitted to the measured points by adjusting three potentiometers representing $y_0 - y_\infty$, y_∞, and K as described in section IV.

IV. Evaluation of Data

a) Theoretical Treatment of Tracer Exchange in a Two Compartment System

It is the aim of most tracer experiments to determine one of the following quantities from the time course of penetration of radioactivity.

1. The time constant K representing the reciprocal of the time interval in which the concentration of the penetrating isotope decreases to $1/e$ (37%) of its original concentration (dimension: 1/sec).

2. The ion flux m representing the number of moles of labelled plus nonlabelled particles of the same species (e.g. $^{39}K + {}^{42}K$, $^{23}Na + {}^{24}Na$, etc.) which penetrates in unit time through unit membrane surface (dimension: moles/cm²/sec).

3. The permeability constant P representing the quotient flux: concentration (dimension: cm/sec).

The ion exchange across the erythrocyte membrane as measured at Donnan equilibrium follows the kinetics of a two compartment system (Fig. 1). Such a system consists of a membrane separating two solutions. It is assumed that the membrane offers resistance to the diffusion of solutes but that its content of the diffusing substance is negligible as compared with the amounts of that substance in each of the adjacent compartments (otherwise the membrane would represent an additional compartment). It is furthermore assumed that only the penetration across the membrane is rate limiting and that no concentration gradients exist inside the two compartments. If such a two compartment system is in the steady state, i.e. if no chemically detectable net movements occur, the time course of disappearance of radioactivity follows a single exponential with time constant K from which m and P can be calculated [Eq. (11), below].

Using the symbols defined in the legend of Fig. 1 the relationship between radioactivity in supernatant and time can be derived as follows:

The change of the amount of radioactivity in compartment a is equal to the difference of the amounts of radioactivity moving from i to a and a to i (conservation of matter):

$$\frac{dy_a}{dt} = -\frac{dy_{ai}}{dt} + \frac{dy_{ia}}{dt} . \qquad (1)$$

Since radioactive and nonradioactive isotopes have indistinguishable properties, the percentage of change of the amounts of labelled and unlabelled substances is equal:

$$\frac{dy_{ai}}{y_a} = \frac{dn_{ai}}{n_a} ; \qquad \frac{dy_{ia}}{y_i} = \frac{dn_{ia}}{n_i} . \qquad (2)$$

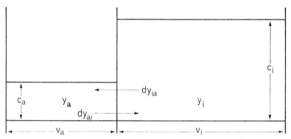

Fig. 1. Two compartment system: y_a amount of radioactivity in compartment a (counts/min). dy_{ai} amount of radioactivity which penetrates from compartment a into compartment i (counts/min). dy_{ia} amount of radioactivity which penetrates from i to a in the same period of time (counts/min). c_a, c_i concentration of the non-radioactive isotope in compartments a and i (Mol/ml). v_a, v_i volumes of compartments a and i (ml). $n_i = c_i \cdot v_i$, $n_a = c_a \cdot v_a$ amount of nonradioactive isotopes in compartments i and a respectively (moles). F membrane surface (cm²)

Introducing the definition of flux

$$m_{ia} = \frac{1}{F} \frac{dn_{ia}}{dt} ; \qquad m_{ai} = \frac{1}{F} \frac{dn_{ai}}{dt} \qquad (3)$$

yields:

$$\frac{dy_{ai}}{dt} = F \cdot \frac{y_a}{n_a} m_{ai} ; \qquad \frac{dy_{ia}}{dt} = F \cdot \frac{y_i}{n_i} m_{ia} . \qquad (4)$$

Inserting (3) and (4) into (1) in the following relation results:

$$\frac{dy_a}{dt} = -F \frac{y_a}{n_a} m_{ai} + F \frac{y_i}{n_i} m_{ia} . \qquad (5)$$

In the steady state the flux from a to i is equal to the flux from i to a:

$$m_{ai} = m_{ia} = m . \qquad (6)$$

In a two compartment system the total radioactivity $\bar{y}$ is equal to the sum of the amounts of radioactivity in compartments i and a:

$$y_i + y_a = \bar{y} . \qquad (7)$$

Combination of (5), (6), and (7) yields:

$$\frac{dy_a}{dt} = mF\left[-\frac{y_a}{n_a} + \frac{\bar{y} - y_a}{n_i} \right] = mF\left[\frac{\bar{y}}{n_i} - \frac{n_i + n_a}{n_i\,n_a} y_a \right]. \tag{8}$$

With the abbreviations

$$\frac{\bar{y}}{n_i} = A \quad \text{and} \quad \frac{n_i + n_a}{n_i\,n_a} = B \tag{9}$$

one obtains:

$$\frac{dy_a}{A - By_a} = mF\,dt. \tag{10}$$

Integration with the initial condition $y_a = \bar{y}$ at $t = 0$ (i.e. at zero time the total amount of radioactivity is present in compartment a) yields:

$$\frac{A - By_a}{A - B\bar{y}} = e^{-mBFt}.$$

For

$$t = \infty,\ e^{-mBRt} = 0,\ \text{and}\ y_a = y_{a\infty}$$

hence: $y_{a\infty} = A/B$ and

$$\boxed{\frac{y_{a\infty} - y_a}{y_{a\infty} - \bar{y}} = e^{-Kt}} \tag{11}$$

where

$$B = \frac{c_i v_i + c_a v_a}{c_i v_i c_i v_a}, \quad \text{and} \quad K = mBF. \tag{12}$$

The permeability constant P is obtained by dividing m by c_a. If the cell surface area is unknown (as is the case in our experiments) one has to be satisfied with the determination of $m_F = m_i F$ and $P_F = m_F/c_a$.

b) Numerical Evaluation of Data

1. Determination of $\bar{y}$, y_∞, and K by means of an analog computer.

An e-function is generated by patching the following circuit (Fig. 2):

The output of amplifier A 02 yields the desired e-function which can be visualized on the screen of the oscilloscope if the computer is operated in the repetitive mode. After completion of the standardization procedure described below the generated curve is fitted to the measured points by adjusting potentiometers P 01, P 02, and P 03 which represent $(y - y_\infty)$, y_∞, and K respectively.

For standardization the three potentiometers are first set to zero by turning the 'position' knob on the oscilloscope such that the beam of the cathode tube is shifted on the y axis until it coincides with the abscissa of the graph. Subsequently P 01 is set to $+10$ V and, by changing the sensitivity control of the oscilloscope, the beam is lifted up to exactly 8 cm above the abscissa. P 01 is then readjusted to zero after which the

beam should return to its original base line. If not, the described procedure has to be repeated until this is achieved. Since P 01 and P 02
attenuate the same reference voltage it is not necessary to standardize
separately P 02. P 03 is standardized by setting P 01 and P 02 to -10 V
and ± 0 V respectively and by adjusting the time scale of the computer

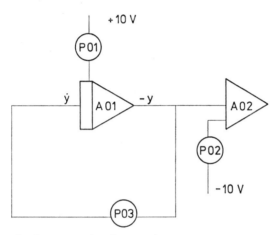

Fig. 2. Analog circuit representing the equation

$$y = (y_0 - y_\infty) e^{-Kt} + y_\infty$$

A 01, A 02, amplifiers; P 01, P 02, P 03, potentiometers; $+$ 10 V, $-$ 10 V current
sources; A 01 is petched together with an integrating network

until the beam passes through the point with the coordinates $y = y_\infty +$
4000 counts/min, $t = 90$ min. Inserting these values in Eq. (11) yields

$$4000 = 8000 \; e^{-90 \, K}$$

and

$$K = \frac{1}{90} \ln \frac{8000}{4000} = 7.71 \cdot 10^{-3}/\text{min} \; .$$

Another reference point is easily obtained for P 03 where $K = 0$ and
$y = 8000$ counts/min. Since a linear relationship must exist between the
readings of P 03 and K the calibration curve is obtained by plotting a
straight line through these points.

The described calibration procedure is somewhat unorthodox and
deviates from ordinary scaling methods. However, it proved to be useful
in cases where the computer is operated by people unfamiliar with
scaling techniques: if measured values of $\bar{y}$, y_∞, or K fall beyond the
calibrated range the outlined methods permit a recalibration without
any further help.

Details concerning any special type of analog computer used for the class work (Telefunken VAT 417 and Pace TR 48) should be described during the course.

Different observers are usually able to reproduce K-values to $\pm 10\%$ or better. It is also possible to fit the data on a digital computer. A during FORTRAN program has been devised by Pring (1968).

2. *Calculation of flux and permeability constant.*

According to Eq. (12) the quantity B has to be determined in order to calculate flux and permeability constant. This quantity can be in-

Table 1

(1)	(2)	(3)	(4)
Concentration in medium added to the cells at start of preincubation period	$\bar{y}$ Radioactivity in supernatant at $t = 0$	$\bar{y}_\infty$ Radioactivity in supernatant at $t = \infty$	$\dfrac{y_\infty}{y_0}$

(5)	(6)	(7)	(8)
Volume of supernatant of 1 ml suspension $1 - Hct$	Dilution of added medium by extracellular fluid of cell sediment $1/1.9\,(1 - Hct)$	Concentr. in supernatant immediately after mixing medium and cells $(1) \cdot (6)$	Concentr. in supernatant at end of equilibration period (C_a) $(4) \cdot (7)$

(9)	(10)	(11)	(12)
Amount in supernatent at equilibrium $N_a = C_a V_a$ $= C_a(1 - Hct)$ $(5) \cdot (8)$	Total amount of substance in 1 ml suspension $(1)/1.9$	Amount of substance i inside cells (n_i) $(10) - (9)$	Water content of 1 ml suspension

(13)	(14)	(15)	(16)
Cell water $(12) - (5)$	Conc. inside the cells at end of equilibration period $(11)/(13)$	Distribution ratio c_i/c_a $(14)/(8)$	Donnan ratio of anions $\sqrt{(14)/(8)}$

(17)	(18)	(19)	(20)
$\dfrac{1}{B} = \dfrac{n_i \cdot n_a}{n_i + n_a}$ $\dfrac{(9) \cdot (11)}{(10)}$	K	$m_F = K/B$ $(18) \cdot (17)$	$P_F = \dfrac{m_F}{c_a}$ $(19)/(8)$

ferred from the known concentration of nonradioactive material added to a known weight of cells at the beginning of the preincubation period, and the ratio y_∞/y_0 as determined on the computer. Since the equilibrium positions for the labelled and unlabelled molecules of the test compound used in the experiments are identical, the decrease of radioactivity in the course of the tracer exchange, y_∞/y_0 is equal to the decrease of the concentration of the unlabelled compound initially added to the external solution. The concentration of the latter in the supernatant at zero time of the preincubation period can be calculated by making allowance for the dilution of the added SO_4 or erythritol solution with the extracellular fluid of the erythrocyte sediment. The dilution amounts to $1/1.9 \cdot (1 - Hct)$ whereby the factor 1.9 represents the volume comprised by 1 ml of original medium and 1 gm of cells. Multiplying the concentration of the original medium by this magnitude yields the concentration of the added substance at zero time in the supernatant. If multiplied by y_∞/y_0, one obtains c_a, the equilibrium concentration of the substance under study. Multiplication of this expression by $(1 - Hct)$ yields n_a, the amount of substance in the supernatant at Donnan equilibrium. By subtracting this quantity from the total amount of substance present in 1 ml suspension one obtains n_i, the amount of that substance inside the cell. The latter is converted into the intracellular concentration by division with the cellular water content. From these data, B as well as the Donnan distribution ratio are calculated and, using the values for B and c_a, m_F and P_F are estimated. To facilitate this somewhat involved treatment of data a table is attached which summarizes the outlined steps and which may be used for the evaluation of the experimental data (Table 1).

V. Results

At a given Cl concentration in the medium, the sulfate flux increases with decreasing pH (experiments 1 and 2), at a given pH it increases with decreasing Cl concentration in the medium (experiments 5 and 6). These findings were interpreted in terms of the fixed charge hypothesis (PASSOW, 1964, 1969): at low pH values the dissociation equilibrium of membrane bound amino groups is shifted to the left: $R - NH_3^+ = R - NH_2 + H^+$. Thus, a larger number of diffusible anions is required to maintain electrical neutrality inside the membrane and hence the anion flux increases. Decreasing the Cl concentration reduces Cl/SO_4 competition for the limited number of fixed charges. These effects are independent of the chemical nature of the anion species employed (experiments 7 and 8) and absent if a nonelectrolyte, erythritol, is used. Even though sulfate permeability is passive, the temperature dependence is high (experiments 3 and 4).

VI. Comments

Measuring ion exchange with tracers at equilibrium offers great advantages over measurements of net movements. Net movements are of necessity accompanied by a continuous change of the composition of medium and cell water; during the course of sulfate penetration, chloride and hydroxyl ions leave the cells in exchange for sulfate which leads to concomitant variations of pH and chloride concentration in cells and medium. Since the transfer of sulfate across the membrane is greatly affected by both pH and Cl-concentration the permeability constant continuously varies in the course of SO_4 net movements. Moreover, since two Cl's leave for each SO_4 entering, the cells shrink and hence a bulk water flow is induced which may or may not influence the rate of ion penetration. Ion movements are usually associated with diffusion potentials which in turn affect the diffusion rate. Even if the transmembrane potential could be measured, its influence on transfer across the membrane is difficult to assess, since the driving force is not the overall membrane potential but the potential gradient $d\psi/dx$ which varies in an unknown way from one point to the next in the membrane. These difficulties in the interpretation of transport processes become negligible if tracer fluxes are studied at equilibrium: the composition of the medium as well as the rate constant is time independent and diffusion potentials and water movements can be neglected.

In erythrocytes the advantages associated with measuring tracer fluxes at equilibrium can be easily illustrated. Net movements of sulfate obey second order kinetics whereas tracer fluxes follow first order kinetics. The usual kinetics of net movements are mainly due to the variations of the composition of the medium and hence the resistance of the membrane. The net ion exchange can be described by the equation

$$\frac{d\,SO_4}{dt} = -K^* \,(SO_4 - SO_{4\infty}) \tag{13}$$

whereby in contrast to tracer exchange at equilibrium, K^* is not a constant but, similar to the driving force $(SO_4 - SO_{4\infty})$, a function of the distance of the system from Donnan equilibrium.

$$K^* = K^{**} \,(SO_4 - SO_{4\infty}) \,. \tag{14}$$

In this latter expression K^{**} is a constant.

Obviously at each instant of the ion exchange the term $(SO_4 - SO_{4\infty})$ describes the composition of the medium with respect to Cl, SO_4, and H concentration (Passow, 1964).

Although the advantages of tracer measurements at equilibrium are obvious, their interpretation is not always as simple as in our special example or a few related cases such as the water transfer across the red cell membrane. In many instances the time course of change of radio-activity does not follow a single exponential. In such cases it is usually assumed that more than one compartment exists in the system. In cell populations, cells with differing permeability constants and ion contents may exist beside each other and in addition each cell may be subdivided into several compartments. The mathematics of such systems has been extensively dealt with and a number of pertinent papers is included in the list of references (SHEPPARD, 1962; SHEPPARD and HOUSEHOLDER, 1951).

In this connection it must be emphasized that the evaluation of our results entirely rests on the assumption that a steady state existed throughout the experiment; the chloride determinations at the beginning and the end of the experiment served to ensure that there were no net anion movements. If net movements occur [i.e. if $m_{ia} \neq m_{ai}$, c.f. Eq. (6)] the position of the isotope equilibrium is constantly changing and hence the tracer movements do not follow equilibrium kinetics any more. A very instructive example of the influence of a change of equilibrium position on ^{42}K movements was published by KEYNES and LEWIS. Using nerve fibres of *Carcinus maenas* they demonstrated that ^{42}K uptake passes through a flat maximum while ^{39}K is leaving the cells. Without closer analysis this result could have lead to the erroneous assumption that only part of the intracellular K^+ was exchanged. For the interpretation of isotope experiments it is therefore absolutely necessary to test by chemical analysis whether or not net movements occur.

The percentage change of radioactivity in the medium between beginning and end of a tracer experiment at Donnan equilibrium depends on the position of equilibrium. In erythrocytes suspended in isotonic saline the Donnan ratio is about 0.7 and hence at a hematocrit of about 50% only 30% of the radioactivity can be expected to enter the cells. This change of radioactivity is rather small and high accuracy is required in order to obtain a reliable estimate of K. The statistical error of the radioactivity measurements as well as the error of pipetting has to be kept below 0.5%, and it is necessary to check whether small variations of the geometry of the counting arrangement affect the results. Due to the low ionic strength of the medium, in the experiments described above, the Donnan ratio was shifted to abnormally high values and hence the demands for accuracy were less severe than for cells in ordinary saline or plasma.

Acknowledgments

The analog computer was supplied by the Volkswagen Foundation. Part of the work was supported by the Deutsche Forschungsgemein-schaft.

2*

References

KEYNES, R. D., and P. R. LEWIS: J. Physiol. (Lond.) 113, 73—98 (1951).
PASSOW, H.: Excerpta Medica International Congress, Ser. No. 87, p. 555, 1965.
— In: The red blood cell. BISHOP, C., and D. M. SURGENOR, Eds. New York: Academic Press 1964.
— Progress in biophysics and molecular biology. London: Pergamon Press (1969 in press).
RAKER, J. W., I. M. TAYLOR, U. M. WELLER, and A. B. HASTINGS: J. gen. Physiol. 33, 691 (1950).
SHEPPARD, C. W.: XVIII, Basic principles of tracer method. London: Wiley 1962.
—, and A. S. HOUSEHOLDER: J. appl. Physiol. 22, 510 (1951).
SOLOMON, A. K.: (1) In: COMAR, C. L., and F. BRONNER, Eds., Mineral metabolism, p. 119. New York: Academic Press 1960.

Ion Permeability of Erythrocyte Ghosts

By H. PASSOW

Technical Problem. The cell contents of human erythrocytes are to be replaced by solutions of known composition.

Principle of Method. The cell contents are removed by hypotonic hemolysis. Substances of our own choice are incorporated into the cells by adding them to the hemolyzing fluid prior to or immediately after hemolysis. Subsequently the leaky membranes are incubated at 37 °C for about 60 min. During this period they largely regain their original impermeability for cations, ATP, and other substances. The degree of recovery is critically dependent on the conditions of hemolysis, in particular on the temperature and alkaline earth ion concentration of the hemolyzing fluid.

Special Application. In six experiments the effects of varying the temperature and alkaline earth ion concentration of the hemolyzing fluid on the incorporation of potassium into the ghosts is explored. The cells are hemolyzed in the presence of 4 mmols/l $MgSO_4$ at 0°, 25°, and 37 °C, or at 25 °C in the presence of a complexing agent, ethylene diamine tetraacetic acid (EDTA), and varying concentrations of Mg^{++}. After resealing, potassium and sodium content and the passive net ion movements between ghosts and medium are determined. Further experiments are designed to illustrate the fact that the ghost membrane retains many of the properties of the membranes of intact erythrocytes: Active transport of sodium in ghosts loaded with ATP as a substrate is demonstrated by following net sodium movements against an electrochemical gradient in the presence and absence of a specific inhibitor of the pump, ouabain. The similarity of the properties of the permeability barrier in ghosts and intact erythrocytes is shown by the similarity of their responses to the action of lead; in intact cells as well as in potassium loaded ghosts, lead induces potassium leakage with little concomitant sodium uptake from the medium.

Instruments. Flame photometer, refrigerated centrifuge, ice bath, thermostatically controlled water baths adjusted to 37 °C and 25 °C, liquid scintillation counter, transparent plastic centrifuge tubes (about 40 ml capacity), pipettes, Erlenmeyer flasks (100 ml capacity), volumetric flasks (50 ml and 5 ml capacity respectively), pH meter, stop watch.

Solutions.

a) 200 ml TRIS-buffer 166 mmoles/l, pH 7.4;

b) 240 μCi ^{14}C-sucrose;

c) 50 ml EDTA, 4 mmoles/l, pH 7.4;

d) 50 ml MgSO$_4$, 1 mmole/l;

e) 100 ml ATP, 4 mmoles/l and MgSO$_4$, 4 mmoles/l, pH 7.4;

f) 200 ml MgSO$_4$, 4 mmoles/l;

g) 20 ml KCl, 3.32 moles/l;

h) 20 ml NaCl, 3.32 moles/l;

i) 2.0 l NaCl, 166 mmoles/l and TRIS, 5 mmoles/l;

j) 200 ml NaCl, 40 mmoles/l KCL 20 mmoles/l choline chloride, 106 mmoles/l, and TRIS pH 7.4, 5 mmoles/l;

k) 50 ml NaCl, 83 mmoles/l, KCl, 20 mmoles/l, choline chloride, 63 mmoles/l, and TRIS, 5 mmoles/l, pH 7.4, ouabain 0.01 mg/ml;

l) 100 ml Pb acetate, 0.08 mmoles/l in solution i);

m) 2000 ml choline chloride, 116 mmoles/l, TRIS, 5 mmoles/l, pH 7.4, and LiCl 50 mmoles/l;

n) 100 ml blood;

o) 50 ml 24% trichloracetic acid (TCE);

p) liquid scintillator (400 ml ethanol, 600 ml toluene, 2.0 g PPO, 0.025 g POPOP);

q) standard solutions for flame photometry of K and Na; concentration range 0.1 − 0.6 mmoles/l; use propan gas.

r) solution e) containing onabain, 0.01 mg/ml.

Execution

Preparation of Ghosts. 100 ml blood is collected into an Erlenmeyer flask containing 20 ml 3.8% sodium citrate. The cells are centrifuged at 1600 g for 10 min. The supernatant, including the buffy coat, is removed. After resuspension in 166 mmoles/l TRIS-buffered NaCl (solution i) the cells are centrifuged again. This washing procedure is repeated twice more. The final sediment is resuspended in 166 mmoles/l TRIS, pH 7.4 to give a 50% cell suspension.

3 ml samples of this suspension are pipetted into eight centrifuge tubes (capacity about 40 − 50 ml) and incubated for 5 min at the temperature at which hemolysis is to be performed (0°, 25°, and 37 °C). Subsequently 30 ml of hemolyzing fluid at the same temperature is added. 5 min after hemolysis, isotonicity is restored by the addition of 1.75 ml of 3.32 molar salt solution. Another 5 min later, the suspensions are transferred to a thermostat adjusted to 37 °C and incubated at that temperature for 45 − 60 min. Subsequently the ghosts are washed twice (refrigerated centrifuge, 10 min at 34500 g) in solutions of the same

Table 1

Tube No.	Hemolysis temperature	Hemolyzing fluid	Restore isotonicity by adding	Wash twice in	At zero time resuspend in	Problem
1	0 °C	4 mmols/l MgSO$_4$ (solution f)	1.75 ml KCl (solution g)	NaCl-TRIS (solution i)	Pb-Acetate in NaCl-TRIS (solution l)	Effects of lead on passive K$^+$ and Na$^+$ movements. Tube No. 6 serves as a control without lead
2	0 °C	4 mmols/l MgSO$_4$ + 4 mmols/l ATP (solution e)	0.87 ml KCl (solution g) + 0.87 ml NaCl (solution h)	KCl/NaCl/TRIS choline (solution j)	Solution j	Active transport of K$^+$ and Na$^+$ and its inhibition by ouabain
3	0 °C	solution r (= e + Ouabain)			Solution k (= j + ouabain)	
4	25 °C	4 mmols/l EDTA (solution c)	1.75 ml KCl (solution g)	NaCl-TRIS (solution i)	NaCl-TRIS (solution i)	Effects of a complexing agent and Mg^{++} on K$^+$ retention. See tube No. 7 for effect of 4 mmols/l Mg^{++}
5	25 °C	1 mmol/l MgSO$_4$ (solution d)	1.75 ml KCl (solution g)	NaCl-TRIS (solution i)	NaCl-TRIS (solution i)	
6	0 °C	4 mmols/l MgSO$_4$ (solution f)	1.75 ml KCl (solution g)	NaCl-TRIS (solution i)	NaCl-TRIS (solution i)	Effects of temperature of hemolyzing fluid on K$^+$ incorporation
7	25 °C	4 mmols/l MgSO$_4$ (solution f)	1.75 ml KCl (solution g)	NaCl-TRIS (solution i)	NaCl-TRIS (solution i)	
8	37 °C	4 mmols/l MgSO$_4$ (solution f)	1.75 ml KCl (solution g)	NaCl-TRIS (solution i)	NaCl-TRIS (solution i)	

Hemolysis: add 30.0 ml of hemolyzing fluid to 3.0 ml 50 vol. % cell suspension in isotonic TRIS pH 7.4 (solution i) *5 min later*: restore isotonicity as indicated above, wait for 5 more min, then transfer to 37 °C for 60 min. Subsequently, wash twice in refrigerated centrifuge, remove supernatant, place all tubes into thermostat, 37 °C. Start experiment by adding 30.0 ml of medium indicated in the sixth column of this table. Take samples after 2, 15, 30, 60, and 120 min.

compositions as that of the media in which the ion movements are to be studied, except that no poisons are added (solutions i and j). After the final wash the supernatants are carefully removed and the centrifuge tubes containing the sedimented ghosts are placed into a water bath adjusted to 37 °C. To each tube 30 ml of the respective suspension medium is added (solutions i, j, k, l) and a stop watch is started. After 2, 15, 30, 60, and 120 min, 3.0 ml samples are withdrawn from each suspension and transferred into centrifuge tubes containing 30 ml ice cold isotonic choline chloride-lithium chloride solution (solution m) and centrifuged at 34 500 g for 5 min. The sediments are transferred to 50 ml volumetric flasks and made up to volume with distilled water. K, Na, and Li are determined by means of a flame photometer. A summary of the experimental procedure is presented in Table 1.

If the effect of poisons on permeability is to be studied, it is usually necessary to ascertain that the results are not invalidated by lysis of the ghosts. Red cell ghosts are impermeable to sucrose. Lysis can be followed therefore by preloading the ghosts with ^{14}C-sucrose and measuring the radioactivity in the sediment of a measured volume of the ghost suspension. Loading the ghosts is achieved by adding ^{14}C-sucrose (1 μCi/ml) to the hemolyzing fluid. For ^{14}C determinations additional 3.0 ml samples are withdrawn as described above: The sedimented ghosts are resuspended in 2 ml distilled water and precipitated by 1 ml 24% trichlor acetic acid (TCA); after centrifugation for 5 min the supernatant is transferred into 5.0 ml volumetric flasks and made up to volume with water. A 0.25 ml aliquot is pipetted into 10 ml of scintillation fluid and counted in a liquid scintillation counter. An alternative method for measuring hemolysis consists in comparing the hemoglobin concentration in the supernatant with that in the suspension as a whole, i.e. in cells plus medium. The supernatant should be diluted 1:2.5 and the suspension 1:25 with distilled water containing a trace of some hemolyzing agent. Hb is estimated photometrically at 418 mμ.

Evaluation of Results

Intracellular contents of K and Na are expressed in terms of μmoles/ml of original erythrocyte volume. Assuming that, at zero time, in each experiment 1.5 ml sediment (i. e. the cells of 3.0 ml 50% erythrocyte suspension) was mixed with 30 ml medium to give a final volume of 31.5 ml, the volume of erythrocytes transferred for flame photometry to the 50 ml flasks amounted to 3.0 × 1.5 × hematocrit of sediment/31.5 = 0.143 × hematocrit. For the intracellular ion content per original erythrocyte volume one obtains: 50 × C/0.143 × hematocrit where C represents the concentration in μmol/ml as measured by flame photometry. For the present purpose it may be sufficient to assume a hematocrit of 0.9. For more accurate determinations it would be preferable to determine the cell concentration for each experiment by means of a Coulter Counter and to express the results in terms of μmoles/cell.

Even though the cells are diluted in a sodium free medium before centrifugation, sodium determinations are subject to a considerable error if the sodium of the medium trapped between the sedimented ghosts is neglected. Assuming that no Li enters the ghosts during the short time interval between suspending them in LiCl-choline solution and sedimentation in the centrifuge, Li determinations in the sediment allow an estimate of the amount of trapped Na.

The volume of the trapped fluid can be calculated if the flame photometer reading of 0.2 ml of LiCl-choline medium diluted to 50.0 ml is compared with the Li readings of the unknown samples:

$$V = 0.2 \cdot \frac{\text{reading of unknown sample}}{\text{reading of 0.2 ml washing fluid}}.$$

The amount of extracellular sodium is obtained by multiplying the Na concentration of the supernatant, C'_{Na}, by V. Since total Na amounts to $50 \times C_{Na}$, intracellular Na is equal to $50\,C_{Na} - VC'_{Na}$. This result has to be converted into μmoles/ml of original cell volume as described above. Estimates of extracellular space by this method give results similar to estimates with inulin or sucrose.

Results

The amount of K^+ incorporated into the ghosts after hemolysis varies considerably with the temperature at which hemolysis was performed. The lower the temperature the higher the amount of K^+ retained by the ghosts (tubes No. 6, 7, and 8).

The action of complexing agents like EDTA (or ATP) and of alkaline earth ions on K^+ incorporation depends on the temperature at which hemolysis took place. At 25 °C, the complexing agents prevent incorporation virtually completely (tube No. 4) whereas increasing Mg^{++} concentrations in the hemolyzing fluid promote K^+ retention (tubes No. 5, 7). After hemolysis at 37 °C, there is little if any resealing of the ghosts reagardless of the presence or absence of complexing agents or alkaline earth ions in the hemolyzing fluid. If present during hemolysis at 0° neither complexing agents nor Mg^{++} exert strong effects. After rewarming to 37 °C, the ghosts reseal nearly perfectly, even in the presence of EDTA, and retain their trapped K^+ quite effectively (as well as trapped complexing agents such as ATP or EDTA) after resuspension in NaCl solution.

The remaining experiments (tubes No. 1, 6, 2, and 3) demonstrate that certain well known properties of the pump as well as of the leak of intact red cells are also present in ghosts. After lead poisoning, resealed ghosts loose approximately 90% of their intracellular potassium within about 45 min. There is little concomitant sodium uptake (compare tubes 1 and 6). Resealing of ATP loaded ghosts in solutions containing equal concentrations of NaCl and KCl yields cells which contain more K and much less Na (tube 2) than a control (tube 6) preincubated without added ATP. The presence of ouabain (tube 3) abolishes this effect of active transport.

Comments

The fact that the ghost membrane largely recovers its cation impermeability has been discovered by Teorell (1952) and Straub (1953). The former studied the osmotic behavior of erythrocyte ghosts whereas the latter used them to demonstrate that ATP provides the energy for active cation transport. The present method is based on the work of Hoffman, Tosteson and Whittam (1964) and Lepke and Passow (1968). The use of resealed ghosts proved to be particularly rewarding in studies of active cation transport (Hoffman, 1962; Whittam, 1962; Glynn, 1962; and others). Ghosts prepared according to the method described in this paper are impermeable enough to be useful for studies of passive cation permeability and have been employed in investigations of the nature of fluoride action on potassium and sodium permeability of the red blood cell (Lepke and Passow, 1968).

The resealed ghosts still contain about 5% of their original hemoglobin. They are incapable of converting adenosine, inosine, or glucose into lactic acid. Their properties depend on the osmolarities of the hemolyzing fluid. The permeability barrier is irreversibly destroyed if hemolysis is performed at very low osmolarities. Hemoglobin free ghosts can be prepared by several closely related methods (e.g. Dodge, Mitchell and Hanahan, 1963). These methods usually involve stepwise hemolysis at more than one tonicity. Such membrane preparations are useful for determinations of the chemical composition of the erythrocyte membrane.

So far the erythrocyte seems to be the only cell whose cell contents can be replaced by reversible cytolysis. In the squid axon it is possible to remove the axoplasm by some mechanical device and to perfuse with solutions of known composition. Since its first description (Baker, Hodgkin, Shaw, 1962) this method has been widely used.

Acknowledgments

My thanks are due to Dr. H. Porzig for a trial run of the described experiments. The work was supported by the Deutsche Forschungsgemeinschaft.

References

BAKER, P., A. L. HODGKIN, and T. I. SHAW: J. Physiol. (Lond.) **164**, 330 (1962).
DODGE, J. T., C. M. MITCHELL, and D. J. HANAHAN: Arch. Biochem. **100**, 119 (1963).
GLYNN, I. M.: J. Physiol. (Lond.) **160**, 18 P (1962).
HOFFMAN, J. F.: Circulation **26**, 1201 (1962).
—, D. C. TOSTESON, and R. WHITTAM: Nature (Lond.) **185**, 186 (1960).
LEPKE, S., u. H. PASSOW: Pflüg. Arch. ges. Physiol. **289**, R 14 (1966).
— — J. gen. Physiol. **51**, 3655 (1968).
STRAUB, F. B.: Acta physiol. Acad. Sci. hung. **4**, 235 (1953).
TEORELL, T.: J. gen. Physiol. **35**, 669 (1952).
WHITTAM, R.: Biochem. J. **84**, 110 (1962).

Countertransport in Human Red Blood Cells

By W. WILBRANDT

I. Principle of Counterflow [1, 3—6]

The term counterflow designates a special case of interactions between simultaneous movements of two different substrates of the same carrier transport system. Its occurence has been demonstrated in a number of cases where saturation kinetics and other evidence suggested the participation of carriers in transport processes. In general, the term is used to designate movement of one substrate S out of equilibrium (i.e. uphill), induced and energized by the simultaneous movement of a second substrate R down its concentration gradient. Under these conditions the direction of the induced flow is opposite to that of the inducing substrate. This effect led to the introduction of the term 'countertransport' or 'counterflow'. However, the interaction is not restricted to this special case but also is observed when the substate S moves with finite rate in the same direction as R. Then the interaction leads to slowing of the movement of S.

In a carrier-system counterflow is a consequence of asymmetric competition for carrier sites. If one substrate is present at equal concentrations at both sides of the membrane, the carriers are equally distributed throughout the membrane. No concentration gradients exist and hence no net flow occurs. The addition of a second substrate reduces the number of sites available for the first one. If the second substrate is present at different concentrations at both sides of the membrane, concentration gradients for the loaded carriers are established which lead to uphill movements of the previously equally distributed substrate.

It is possible to derive an equation which describes the rate of transport of one substrate S in the presence of a second substrate R. This equation predicts the occurence of counterflow and its dependence on the saturation of the carrier system with S and R. The derivation rests on the following assumptions: 1. in the membrane a steady state is established, 2. the mobilities of loaded and unloaded carriers are equal, 3. the movement of the loaded carriers CR and CS is slow as compared to the rate of reaction between the substrates R or S and carrier C

(equilibrium between substrate and carrier at phase boundaries). For the rate of net flow of S, v_S, one then obtains:

$$v_S = v_{max} \left(\frac{S_1'}{S_1' + R_1' + 1} - \frac{S_2'}{S_2' + R_2' + 1} \right) \tag{1}$$

where the symbols S' and R' represent relative concentrations of S and R, i.e. S/K_{CS} and R/K_{CR} respectively. K_{CS} and K_{CR} denote the dissociation constants of the respective carrier substrate complexes and the indices 1 and 2 refer to the two membrane surfaces.

Eq. (1) can be rewritten in the form:

$$v_S = v_{max} \frac{S_1'(R_2' + 1) - S_2'(R_1' + 1)}{(S_1' + R_1' + 1)(S_2' + R_2' + 1)}. \tag{1a}$$

This equation shows that, in contradistinction to a situation where $R = 0$, under the condition $R \neq 0$ the rate of transport v_S does not approach zero for $S_1 \rightarrow S_2$. At zero transport of S we obtain

$$\frac{S_1}{S_2} = \frac{R_1' + 1}{R_2' + 1} \quad \text{or} \quad \frac{S_1}{S_2} = \frac{R_1 + K_{CR}}{R_2 + K_{CR}}. \tag{2}$$

The maximum concentration ratio attained by countertransport for the substrate S is therefore independent of K_{CS} but dependent on K_{CR}. The maximum ratio varies between 1 and R_1/R_2, depending on K_{CR}. The higher the affinity of R to the carrier, the higher is the maximum concentration ratio.

II. Principle of Method

Washed human red cells are equilibrated with ^{14}C labeled D-xylose. After equilibration is completed the medium ('equilibrium medium') is separated from the cells ('equilibrium cells') by centrifugation. D-glucose (or another sugar) is dissolved in the supernatant, and cells and medium are mixed again. The resulting suspension is incubated at 30 °C. At suitable time intervals samples are withdrawn and centrifuged rapidly. The radioactivity of the supernatant is determined by means of a liquid scintillation counter. The radioactivity rises rapidly but transiently above the original equilibrium level, indicating uphill movement of labeled D-xylose from the cells to the medium.

III. Execution

Instruments and Materials. Centrifuge (about 6000 g), liquid scintillation counting equipment, analytical balance, Erlenmeyer flasks (100 ml), pipettes 10 ml, 1.0 ml, 0.1 ml, bulb pipettor, centrifuge tubes (3.0 ml).

Solutions.

1. buffered saline: 140 mmoles/l NaCl, 20 mmoles/l Na_2HPO_4, adjusted to pH 7.4;

2. buffered saline (solution 1) containing 16.7 mmoles/l ^{14}C-D-xylose, specific activity 0.1 μC/mg;

3. liquid scintillator (see under III 2);

4. 3.8% sodium citrate.

Chemicals.

D-galactose, D-mannose, D-fructose,
L-arabinose, D-arabinose, D-ribose.

1. Loading of Cells with D-Xylose. 50 ml of human blood is drawn into 10 ml 3.8% sodium citrate solution. The mixture is centrifuged, the supernatant, including the buffy coat, is sucked off, and the cells are washed three times in buffered saline (140 mM NaCl, 20 mM Na_2HPO_4 adjusted to pH 7.4). 18 ml of washed and packed cells are resuspended in 36 ml buffered saline containing 16.7 mM ^{14}C-D-xylose (specific activity 0.1 μc/mg) to give suspension I. After mixing cells and medium suspension I is incubated at 30 °C for at least 2 h. At the end of this period D-xylose is equally distributed between cells and medium.

2. Induction of Counter Transport. 15 and 5 min before inducing counter flow by the addition of suitable substrates, about 0.8 ml of suspension I is centrifuged. 0.1 ml of the supernatant is pipetted into scintillation counting vials containing 10 ml liquid scintillator (4 g PPO* + 0.1 g POPOP** per 1 toluene). The solutes are mixed and the vials closed.

Immediately after taking the second sample the rest of suspension I is centrifuged and the equilibrium medium is collected. 540 mg D-glucose (or D-galactose, D-mannose, D-fructose, or 450 mg L-arabinose, D-arabinose, or D-ribose) are dissolved in exactly 10.0 ml of equilibrium medium. The resulting solution is mixed with 10 ml equilibrium cells to give suspension II (= zero time of counter flow experiment) and incubated at 30 °C. After 1, 3, 6, 9, 15, 30, 60, 90 min, samples of approximately 0.8 ml are rapidly centrifuged and in each instance 0.1 ml of the supernatant is pipetted into counting vials and treated as described above.

Radioactivity is counted in a liquid scintillation counter and the results are plotted against time. It is advisable to make all radioactivity determinations in duplicate.

Each pair of students is able to carry out two experiments with two different sugars.

---

* 2.5-diphenyloxazole.
** 2-p-phenylene bis (5-phenyloxazole).

Results

Fig. 1 demonstrates experimental results obtained with a number of penetrating sugars (D-glucose, D-galactose, D-arabinose, D-ribose, D-fructose) and a nonpenetrating solute, mannitol, which does not interact with the sugar transport system. The radioactivity of the medium measured before the addition of the inducing sugar varied actually between 25100 and 27900 cpm. Counter transport (i.e. the transient increase in the radioactivity of the medium) is most pronounced with D-glucose, whereas D-galactose and L-arabinose are less effective and D-ribose yields only dubious results. The curve obtained with D-fructose is indistinguishable from that with mannitol.

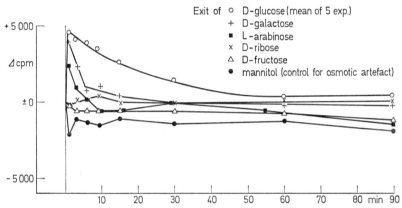

Fig. 1. Counterflow of previously equilibrated ^{14}C-D-Xylose, induced by addition of various sugars to the medium, in red cells. Ordinate: change of radioactivity in the medium, abscissa: time. The inset should read entry rather than exit

Discussion

1. Osmotic Artifacts. Induction of counter flow by adding a substrate to an equilibrium medium is of necessity associated with the establishment of an osmotic imbalance. Under the above experimental conditions, osmotic water shifts are most pronounced at zero time. They have two consequences: 1. a dilution of the radioactivity of the medium and 2. a concomitant increase of intracellular radioactivity. This induces a movement of radioactivity from the cells into the medium. However, this movement can never lead to radioactivity levels of the medium exceeding the initial one. Therefore, under the chosen experimental conditions, an osmotic artifact could not mimic counter transport since osmotic artifacts are associated with a transitory decrease of external radioactivity in contradistinction to the transient increase induced by counter flow. The magnitude of the osmotic artifact superimposed on the counter flow can be inferred from the control experiment where the nonpenetrating, nontransported solute mannitol was added instead of a penetrating sugar.

2. The Maximum Activity Ratio Attained by Counter Flow. The maximum ratio between external and intracellular radioactivity that is attained in the course of a counter flow experiment can be calculated if the ratio between intracellular and extracellular water volume is known. Under the chosen experimental conditions, this volume ratio was approximately 1:3,5 (assuming 10% of medium trapped between the packed cells, and 66% of water in the total cell volume and allowing for osmotic water shift). In the above experiments, only the radioactivity of the medium was measured. From its maximal value a_o and the final radioactivity as measured after reestablishment of equilibrium, a_e (taken as 26500 cpm) the activity ratio can be calculated (the subscripts i and o refer to cell water and medium respectively):

$$\frac{a_o}{a_i} = \frac{a_o}{4,5a_e - 3,5a_o} .$$

This equation follows if one takes into account that the total radioactivity of suspension II is constant throughout the experiment.

From the data presented in the figure the following ratios were calculated:

D-glucose	2.94
D-galactose	2.24
L-arabinose	1.48
D-ribose	1.15
D-fructose	1.0
D-mannitol	1.0

As expected [cf. p. 29, Eq. (2)], the values for the various sugars follow the order of their affinities to the carrier system of the cells. This is seen by comparing them with their Michaelis constants (taken from a graphical presentation of data obtained by LeFèvre and Marshall [2])

D-glucose	8 mM
D-galactose	40 mM
L-arabinose	130 mM
D-ribose	180 mM .

In principle, if the distribution of the inducing sugar at the peak of counter flow were known it should be possible to calculate Michaelis constants for the various sugars from the measured radioactivity ratios. However, since this is not the case and the maximum of the radioactivity of the medium was not accurately determined in the present experiments, one has to be satisfied with the demonstrated qualitative agreement.

References

1. CIRILLO, V. P.: Sugar transport in microorganisms. Ann. Rev. Microbiol. 15, 197—218 (1961).
2. LeFéVRE, P. G., and J. K. MARSHALL: Conformational specificity in a biological sugar transport system. Amer. J. Physiol. 194, 333—337 (1958).
3. PARK, C. R., R. L. POST, C. F. KALMAN, J. H. WRIGHT, JR., L. H. JOHNSON, and H. E. MORGAN: The transport of glucose and other sugars across cell membrane transports involving insulin. Ciba Coll. Endocrin. 9, 240—260 (1956).
4. ROSENBERG, T., and W. WILBRANDT: Uphill transport induced by counterflow. J. gen. Physiol. 41, 289—296 (1957).
5. WIDDAS, W. F.: Inability of diffusion to account for placental glucose transfer in the sheep and consideration of the kinetics of a possible carrier transfer. J. Physiol. (Lond.) 118, 23—39 (1952).
6. WILBRANDT, W., and T. ROSENBERG: The concept of carrier transport and its corollaries in pharmacology. Pharmacol. Rev. 13, 109—183 (1961).

Kinetics of Cation Transport in Yeast

By W. McD. Armstrong and A. Rothstein

I. Introduction

a) **Technical Problem.** The interactions of Rb⁺ and Cs⁺ with the cation transporting system in baker's yeast will be measured under selected experimental conditions using ^{86}Rb and ^{137}Cs as tracers. The data obtained will be analyzed in terms of a kinetic model based on the Michaelis-Menten treatment of enzyme reactions [1].

b) **General Principles of the Method.**

1. Carrier theory and Michaelis-Menten kinetics [2, 3]: The kinetics of transport of a wide variety of physiologically important substances (e.g. sugars, amino acids, and ions) across biological membranes have in many instances been successfully interpreted in terms of "carrier" mechanisms. The basic assumptions involved in this interpretation are 1. reversible binding of the transported substance (substrate) with a component of the membrane (the carrier), 2. movement of the resulting complex from one side of the membrane to the other, and 3. splitting or dissociation of the carrier-substrate complex at the opposite side of the membrane from that at which it is formed. It is further supposed that the carrier, either free or combined, moves back and forth across the membrane in a cyclic fashion. Thus the transfer of relatively large amounts of material can be effected by a comparatively small number of carrier molecules. Schematically, this sequence of steps may be formulated as shown in Fig. 1.

The assumption of carrier mechanisms enables many of the observed kinetic characteristics of biological transport processes to be explained in terms of models which, though undoubtedly oversimplifications of a complex situation, have proved extremely useful in practice. These characteristics include saturation, specificity of transport, and competition. In terms of carrier theory, the general equation for the net rate of transport, V, of a substrate from side 1 to side 2 of a biological membrane may be written

$$V = D(CS_1 - CS_2) \qquad (1)$$

where D is the diffusion coefficient, within the membrane, of the carrier-substrate complex CS (assumed to be the same as the diffusion coefficient for the free carrier) and CS_1 and CS_2 are the concentrations of the carrier-substrate complex on sides 1 and 2 of the membrane.

In practice, CS_1 and CS_2 cannot usually be determined directly and it is necessary to transform Eq. (1) into a kinetic equation which gives V in terms of quantities which can be measured experimentally. The equation

$$V = DC_t \left(\frac{S_1}{S_1 + K_{m_1}} - \frac{S_2}{S_2 + K_{m_2}} \right) \tag{2}$$

is a general equation of this kind. Here S_1 and S_2 are the substrate concentrations on sides 1 and 2 of the membrane, C_t is the total amount of carrier available for transport, and K_{m_1} and K_{m_2} are the Michaelis constants on sides 1 and 2 of the membrane. It is evident that the two terms within the brackets tend towards the limits 1 and 0 respectively as S_1 and S_2 become very large or very small. Hence the maximum possible rate of transport, V_m, is DC_t, i.e. beyond a certain substrate concentration the rate cannot be increased by further increasing the amount of substrate available (saturation). K_m is numerically equal to the substrate

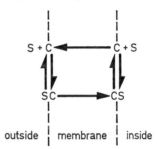

Fig. 1. Schematic representation of a simple carrier operating cyclically across a cell membrane

concentration when $V = 1/2\ V_m$. If it is assumed that, of the three steps involved in substrate transfer; binding, transmembrane movement, and dissociation, the second is rate limiting (i.e. D is smaller than the rate constants for binding and dissociation) K_m may be taken as being approximately equal to K_s, the mass law dissociation constant for the formation of the carrier-substrate complex CS. Hence, the quantity $1/K_m$ may be taken as an approximate measure of the affinity of the substrate for the carrier. Similarly, the relative affinities of a series of substrates which combine with the same carrier can be expressed in terms of their K_m values.

When the second term within the brackets in Eq. (1) approaches zero, the rate equation may be written

$$V = V_m \cdot \frac{(S)}{(S) + K_m} . \tag{3}$$

This is the classical Michaelis-Menten equation of enzyme kinetics [1]. Under these conditions V will depend on the concentration of CS, and the transport kinetics will approximate closely to the Michaelis-Menten

3*

case. In practice, biological transport phenomena are frequently found to obey the Michaelis-Menten equation. Among the conditions under which this occurs one may mention 1. measurements of unidirectional movements of substrate e.g. measurement of the initial rate of penetration of an external substrate into a cell ($S_2 \approx 0$), 2. $K_{m2} \gg K_{m1}$ or vice versa (this can occur if the carrier is altered by metabolic reactions on one side of the membrane and is frequently invoked as a model for "uphill" or active transport. Conversely, it will be evident from Eq. (1) that if $K_{m1} = K_{m2}$ carrier mediated transport can lead only to equilibration of S across the membrane), 3. continuous rapid metabolic removal of S in the interior of a cell.

Transport phenomena which follow Michaelis-Menten kinetics are susceptible to very simple analysis in terms of the parameters K_m and V_m. A useful method for determining these parameters is that of LINE-WEAVER and BURK [4]. In this method Eq. (3) is rearranged to give

$$\frac{1}{V} = \frac{1}{V} + \frac{K_m}{V_{m(S)}} . \tag{4}$$

A plot of $1/V$ against $1/(S)$ then yields a straight line with a Y-intercept equal to $1/V_m$ and a slope equal to K_m/V_m.

2. *Kinetics of inhibition and competition*: Michaelis-Menten kinetics also provide a useful approach to the study of inhibition of transport and of competition between an array of substrates for a common transport mechanism. Two cases are of particular interest, competitive inhibition or competition in which an inhibitor or a competing substrate can combine with the carrier (or the specific site on the carrier) which is implicated in the transport of a given substrate* thereby preventing the combination of the substrate with this site, and non-competitive inhibition in which the combination of an inhibitor with a site other than the transporting site results in inhibition of transport. With a purely competitive inhibitor it is assumed that the rate of transfer across the membrane and of dissociation of the complex CS is not affected. Thus, as in the absence of inhibition, V depends on the concentration of CS. The formation of the complex CI (where I is the inhibitor or competitor) reduces the amount of CS formed and, consequently, diminishes V. The amounts of CS and CI formed will depend on the relative concentrations of substrate and inhibitor and on the relative affinity of the carrier for each of these.

---

* One can distinguish between an inhibitor which combines with the transporting site, thereby preventing the transfer of substrate across the membrane, but which is not itself transported, and a competitor (competing substrate) which combines with the same site but is transported. However, in terms of their effects on substrate transfer, both give rise to the same formal kinetic pattern.

Thus, in the presence of a fixed amount of inhibitor, the degree of inhibition observed will depend on the substrate concentration (S) and on K_s. When (S) is sufficiently high the inhibition will be completely reversed. In non-competitive inhibition, on the other hand, the formation of CS is supposed to be unaffected by the inhibitor but its rate of transfer across the membrane and/or dissociation is reduced. In this case the degree of inhibition depends only on the inhibitor concentration (I) and on K_i (the dissociation constant for the complex CI), and is not affected by the substrate concentration.

In terms of the parameters K_m and V_m, competitive inhibition is characterized by an increase in the apparent K_m for the substrate, V_m remaining unchanged. Non-competitive inhibition results in a decrease in V_m without any apparent change in K_m. Writing α for the change in K_m and β for the change in V_m caused by a given inhibitor, and assuming equilibrium conditions (i.e. that the complexes CS and CI are in equilibrium with their components and that $K_m \approx K_s$), a general equation for the rate of transport of a substrate in the presence of a reversible inhibitor can be derived [3]. This equation has the form

$$V_i = V_m \, \frac{(S) \, [\alpha \, K_i + \beta \, (I)]}{(S) \cdot (I) + \alpha [K_i \cdot (S) + K_m \cdot (I) + K_m \cdot K_i]} . \tag{5}$$

From this equation, using appropriate values of α and β, various specific types of inhibition can be characterized. Four types will be considered herein:

a. *Complete competitive inhibition*: The inhibitor does not affect the rate of transport of the carrier-substrate complex across the membrane or its rate of dissocation, but can, in sufficiently high concentrations, completely prevent the combination of substrate with carrier $(\alpha = \infty : \beta = 1)$. In this situation the rate equation becomes

$$V_i = V_m \cdot \frac{(S)}{(S) + K_m \, [1 + (I)/K_i]} \tag{6}$$

or, in terms of the double reciprocal formulation of Lineweaver and Burk [4]

$$\frac{1}{V_i} = \frac{1}{V_m} + \frac{K_m}{V_m \cdot (S)} \left[1 + \frac{(I)}{K_i}\right] . \tag{7}$$

It is seen that, for competitive inhibition, a plot of $1/V_i$ against $1/(S)$ gives a straight line with the same Y-intercept, $\dfrac{1}{V_m}$, as the control without inhibitor but with a slope which is increased by the factor $[1 + (I)/K_i]$.

b. *Complete non-competitive inhibition*: The inhibitor does not affect the combination of substrate with carrier but can, in sufficiently high concentrations, completely prevent the transfer of the carrier-substrate complex across the membrane and/or its dissociation ($\alpha = 1 : \beta = 0$). The appropriate rate equations are

$$V_i = V_m \, \frac{K_i}{(I) + K_i} \cdot \frac{(S)}{(S) + K_m} \tag{8}$$

and

$$\frac{1}{V_i} = \frac{1}{V_m} \left[1 + \frac{(I)}{K_i} \right] + \frac{K_m}{V_m \cdot (S)} \left[1 + \frac{(I)}{K_i} \right]. \tag{9}$$

In this case a plot of $1/V_i$ vs. $1/(S)$ gives a straight line in which both the Y-intercept and the slope are increased by the factor $[1 + (I)/K_i]$ relative to the control without inhibitor.

c. *Partial competitive inhibition*: The inhibitor reduces the affinity of the carrier for the substrate but does not affect the rate of transport of breakdown of the carrier-substrate complex ($\infty > \alpha > 1 : \beta = 1$). For this type of inhibitor

$$V_i = V_m \, \frac{(S)}{(S) + K_m \, \alpha [(I) + K_i]/[(I) + \alpha \, K_i]} \tag{10}$$

and

$$\frac{1}{V_i} = \frac{1}{V_m} + \frac{K_m}{V_m \cdot (S)} \cdot \frac{\alpha \, [(I) + K_i]}{(I) + \alpha \, K_i}. \tag{11}$$

d. *Partial non-competitive inhibition*: The formation of the carrier-substrate complex is not affected. The rate of transport and/or dissociation of the complex is reduced but not completely inhibited ($\alpha = 1 : 0 < \beta < 1$). In this case

$$V_i = V_m \cdot \frac{\beta \, (I) + K_i}{(I) + K_i} \cdot \frac{(S)}{(S) + K_m} \tag{12}$$

and

$$\frac{1}{V_i} = \frac{1}{V_m} \left[\frac{(I) + K_i}{\beta \, (I) + K_i} \right] + \frac{K_m}{V_m \cdot (S)} \left[\frac{(I) + K_i}{\beta \, (I) + K_i} \right]. \tag{13}$$

In practice, double-reciprocal plots alone are usually not sufficient to characterize inhibitions of the partial competitive or non-competitive type and other methods of graphical analysis may be necessary [3]. It should be noted that the above treatment can be extended to include activation effects as well as inhibition by a suitable choice of values for α and β.

If, for a given inhibitor, the relation between substrate concentration and rate of transport is studied at several different inhibitor concentrations, the double reciprocal plot of Lineweaver and Burk gives a series of straight lines whose relative slopes (and intercepts in the case of non-competitive inhibition) are determined by the concentrations of inhibitor used. A method which fits the data for different inhibitor concentrations to a single curve and which is of considerable practical utility in certain circumstances (see "evaluation of results" below) is that introduced by

Hunter and Downs [5]. For complete competitive inhibition the Hunter and Downs equation is

$$(I) \cdot \frac{a}{1-a} = K_i + \frac{K_i \cdot (S)}{K_m} \tag{14}$$

where a is the fractional activity $(a = V_i/V_o$ where V_i is the rate in the presence of inhibitor, and V_o is the rate, at the same substrate concentration, in the absence of inhibitor). Plotting $(I) \cdot \frac{a}{1-a}$ against (S) one obtains a straight line with an intercept equal to K_i and a slope equal to K_i/K_m.

For complete non-competitive inhibition the Hunter and Downs equations takes the form

$$(I) \cdot \frac{a}{1-a} = K_i . \tag{15}$$

In this case a plot of $(I) \cdot \frac{a}{1-a}$ vs. (S) gives a straight line parallel to the X-axis and having a Y-intercept equal to K_i.

With partial inhibitions of the types discussed above, the Hunter and Downs plots are more complex. With partial non-competitive inhibitors a plot of $(I) \cdot \frac{a}{a-1}$ vs. (S) gives a straight line parallel to the X-axis but the value of the Y-intercept depends on β as well as on K_i. With partial competitive inhibitors both the slope and the intercept depend on the value of α as well as on K_i and K_m [3].

3. *Cation uptake in baker's yeast (Saccharomyces cerevisiae)*: Apart from its ready availability in quantity, baker's yeast offers a number of advantages as a system in which to study the kinetics of cation transport. Structurally, yeast cells are similar to plant cells in that they possess an outer wall of considerable mechanical strength which enables them to withstand large differences in osmotic pressure between the cytoplasm and the external medium. The internal osmolality of the resting yeast cell is about $550 - 600$ m osm [6] although the cells normally exist in a highly diluted medium. Thus, in contrast to animal cells and microorganisms such as E. coli which exist in a saline environment, the passive (leak) permeability of yeast to cations and anions is very low. When supplied with a source of metabolic energy such as glucose, the cells can rapidly take up large quantities of K^+ and other alkali metal ions. Uptake occurs by means of a cation exchange mechanism in which one H^+ ion, produced as a result of metabolic processes, is ejected from the cell for each alkali metal ion taken up [7, 8].

Experimentally, the kinetics of cation uptake by yeast cells can be determined by relatively simple methods. Several factors contribute to this. Among these may be mentioned the facts, already noted above, that in these cells cation uptake is uncomplicated by associated movements of anions and that outward leaks of cations from the cell are negligible compared to the amounts taken up, at least in experiments of relatively short duration (5 − 15 min at room temperature). Further, it has been found [9] that, during the first few minutes following its onset, cation uptake is a linear function of time. Thus, the initial rate of uptake of a given cation can be determined by a allowing the cells to ferment glucose in the presence of the ion, labelled with a radioactive isotope, for a measured period of 5 − 15 min, removing the supernatant fluid by centrifugation or filtration, washing the cells with ice cold water to remove extracellular label, and assaying the amount of radioactivity in the cells.

Because of the relative simplicity of the experimental techniques involved, the kinetics of cation uptake by actively fermenting yeast have been rather extensively investigated. Some of the results obtained in these studies may be briefly mentioned. When the external pH is near neutrality, the initial rates of uptake of each of the alkali metal ions can be fitted by an equation of the form of Eq. (3) [10, 11, 12]. Furthermore, with certain pairs of cations, the uptake of one is inhibited in a competitive manner by the other. Thus, K^+ is a competitive inhibitor of the uptake of Na^+ [10], Rb^+, and Cs^+ [13], and H^+ is a competitive inhibitor, under certain circumstances, for K^+ and the other alkali metal ions [12, 14]. Such studies have been interpreted in terms of a model in which a single carrier serves to transport H^+ and the alkali metal ions into the cell [10]. The array of relative affinities of these ions for the carrier is, $H^+ > K^+ > Rb^+ > Cs^+ > Na^+ > Li^+$ [10, 12, 14].

Recent studies [12, 13] have shown however that the mutual interactions of pairs of cations with the cation transporting system in yeast cannot, in all circumstances, adequately be explained in terms of a simple competition model based on a single binding site. An investigation of the effect of external pH on cation uptake [12] revealed the existence of three kinetic regions between pH 3 and pH 8. Between pH 6 and pH 8, there was no discernible effect of H^+ on the uptake of alkali metal ions. Between pH 4 and pH 6 an inhibitory effect of H^+ was detected which, at relatively low concentrations of the transported ion, had the kinetic characteristics of a partial non-competitive inhibition i.e. a reduction in the apparent V_m for uptake with relatively little change in the apparent K_m. Below pH 4 the remaining cation uptake was in each case competitively inhibited by H^+. Both effects of H^+ could be overcome by increasing the external concentration of the transported ion to sufficiently high levels, but the concentrations required to overcome the apparently noncompetitive effect were, in all cases, considerably higher than the concentrations required to reverse the direct competitive inhibition. Also, it was found that the apparently non-competitive effect of H^+ was greater for the other alkali metal ions than for K^+, leading to an enhanced discrimination of the transport system in favor of K^+ at low pH. These results were interpreted as indicating that H^+ and other cations can combine with two separate sites in the cation transporting system, the transport site itself and a second non-transporting (modifier) site which can influence the maximal rate of transport. Each site has a different specificity for H^+ and the alkali metal ions.

In a subsequent study [13] the effect of K^+ on the uptake of each of the alkali metal ions and, in turn, the influence of these, and also of Mg^{++} and Ca^{++}, on the uptake of K^+ was investigated. In these experiments, effects of H^+ were virtually excluded by working at pH 8. It was found that K^+ competitively inhibits the uptake of each of the other alkali metal ions and that Rb^+ competitively inhibits K^+ uptake. Li^+ Na^+, and Cs^+ act like H^+. At relatively low concentration they partially inhibit the uptake of K^+ in a manner which, at low external K^+ concentrations, shows kinetic characteristics similar to those described above for partial non-competitive inhibitors. At higher concentrations, Li^+, Na^+, and Cs^+ competitively inhibit the residual uptake of K^+. With Cs^+ it is possible, by chosing appropriate experimental conditions, to achieve virtually complete separation of the two inhibitory effects. With Li^+ and Na^+ there is a considerable overlap and, in almost all circumstances, the total observed inhibition is a composite of both effects. As was found previously with H^+ [12], both there effects can be' reversed by high external K^+ concentrations, the direct competitive inhibition being in all cases the more easily reversed of the two.

Mg^{++} and Ca^{++} in relatively low concentrations ($1 - 2$ mM) inhibit K^+ uptake in an apparently non-competitive manner although their affinities for the cation carrier are extremely low [7, 10]. Furthermore, it was found that when inhibiting cations were present in concentrations that produced an apparently non-competitive effect only, they were not taken up in quantity by the cells. During competitive inhibition, the competing cation is taken up in amounts corresponding to the predictions of simple competition theory.

On the basis of these studies a kinetic model for cation uptake in yeast has been proposed [13]. This model is illustrated schematically in Fig. 2. In this figure, C and M are two independent binding sites for cations, C being the carrier or transporting site and M a modifier site which is not a transporting site. C and M have affinity arrays for cations which, though qualitatively similar, are quantitatively different [13]. H^+, K^+ and the other alkali metal ions can combine with both C and M. Mg^{++} and Ca^{++}, except at extremely high concentrations, combine only with M. When C alone or both C and M are occupied by the same ion, S, uptake follows simple Michaelis-Menten kinetics (pathway 1). Occupation of C by a competing species, I, results in competitive inhibition of the uptake of S with I also being taken up in proportion to the fraction of the total number of C sites it occupies (pathway 3). Occupation of M by I usually results in a reduction of the V_m for S. This reduction occurs without any significant uptake of I and can be expressed in terms of the empirical kinetic constant β [Eq. (5)]. Thus $V_m = k_3 C_t$ (where C_t is the total number of C sites available) and $V_{m'} = \beta k_3 C_t$ when all available M sites are occupied by I (pathway 2, Fig. 2). For the interactions of the

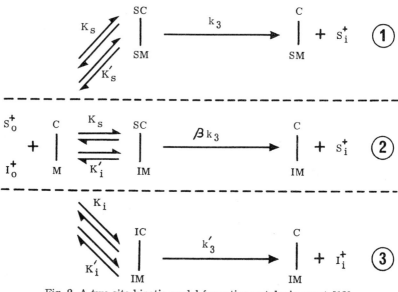

Fig. 2. A two site kinetic model for cation uptake in yeast [13]

various cations studied with the cation transporting system in yeast, β was found to have values between 0 and 1 [12, 13].

On the basis of this model, the interactions of a given cation with the cation transporting system of the yeast cell can be expressed in terms of a set of kinetic parameters. These are K_m, V_m, β, and K_i (the mass law dissociation constant for the combination of the ion with the modifier site M). From the values of K_m and K_i for different ions, affinity arrays for the transporting site and the modifier site can be derived [13]. The present experiment illustrates the determination of the relevant kinetic parameters for two alkali metal cations, Rb^+ and Cs^+. Rb^+, rather than K^+, is chosen for two reasons. First, it generally resembles K^+ closely in its behavior in biological systems and, specifically, the transport system in yeast handles K^+ and Rb^+ in almost identical fashion. Second, the technical difficulties involved in using the short lived isotope ^{42}K can be avoided by using ^{86}Rb. Cs$^+$ is used as the second ion because, as already pointed out, of all the alkali metal ions it gives the clearest distinction between the kinetic behavior of the two sites implicated in cation transport. Also, as the case of Rb$^+$, a suitable radioisotope of Cs$^+$ is available.

c) **Other Applications.** In principle, the method of kinetic analysis illustrated in this experiment is capable of rather wide and useful application to biological transport systems showing saturation kinetics [2]. The type of model demonstrated is becoming increasingly important in the characterization of the control systems that regulate enzyme activity (allosteric systems) [15]. The model also provides useful insights into some of the effects of pH on bilogical transport systems [12, 14].

d) **Equipment, Solutions etc.**

1. General class equipment
a. Scaler, pulse height analyzer, crystal probe and autochanger for planchettes, or, alternatively, an auto-gamma well scintillation spectrometer.
b. 10 ml burettes (self filling) with 0.1 ml graduations (for delivery of radioactive and non-radioactive salt solutions and tris buffer).

2. Equipment for individual working groups
a. Disposable planchettes (to hold 25 mm filter discs).
b. Millipore filters (25 mm diameter; mean pore diameter 1.2 μ).
c. Scintered holders for 25 mm millipore filters with matching clamps.
d. Suction flasks (125 ml) fitted with stoppers containing holes to fit the outlet tube of the filter holders.
e. Suction pumps and tubing.
f. Test tubes (about 20 ml capacity).
g. Automatic burettes to deliver 1 ml or 1 ml pipettes (for delivery of glucose solution).
h. 2 ml pipettes (for delivery of yeast suspension).
i. Wash bottles (plastic squeeze type) containing ice cold water.
j. Glass eye droppers.

3. Solutions
a. 0.1 M tris (trishydroxy-methyl-aminomethane) buffer (adjusted to pH 7 with HCl).
b. The following non-radioactive solutions made up in 0.1 m tris buffer (pH 7): 1 M glucose, 10 mM CsCl and 20 mM RbCl.
c. The following radioactive solutions made in 0.1 M tris buffer (pH 7): Cs*Cl — 50 mM CsCl containing approximately 1 μc ^{137}Cs/ml, Rb*Cl (10 mM) — 10 mM RbCl containing approximately 0.5 μc ^{86}Rb/ml, Rb*Cl (100 mM) — 100 mM RbCl containing approximately 5 μc ^{86}Rb/ml.

d. 5% glycerol in water.

e. 20% sucrose in water.

4. Yeast (saccharomyces cerevisiae)

a. Preparation of starved yeast:

1) Commercial baker's yeast is used in this experiment. Fresh yeast may be used directly following aeration and washing as follows: Roughly weigh a suitable quantity of yeast (about 1 g for each pair of students) and suspend it in about 10 volumes of distilled water. Pass a continuous stream of air or oxygen through the suspension for about 12 — 16 h. Centrifuge and discard the supernatant. Wash twice by resuspending the cells in about 10 volumes of distilled water and centrifuging. Store in a refrigerator in a well stoppered container. This should be done on the day preceding the experiment.

2) If time permits, it is preferable to use a yeast culture grown in the laboratory. Grow a broth culture of the yeast for 24 h in the medium described below. Inoculate the total culture into 1 l of fresh medium. Allow to grow for 24 h at room temperature with continuous aeration. Transfer this liter of material into 4 l of culture medium and allow growth to continue with aeration for a further 24 h. Harvest the cells by centrifugation. Wash three times in distilled water, suspend the cells in about 10 volumes of distilled water and aerate for 12 — 16 h. Centrifuge, wash twice with distilled water and store in a refrigerator. This should provide sufficient yeast for 60 students.

3) Culture medium. — This should contain 1% yeast extract, 2% peptone, 2% glucose or sucrose, 0.1% $KHPO_4$, 0.12% $(NH_4)_2SO_4$, and 0.01% $MgCl_2$ and should be adjusted to pH 5 with HCl or NaOH. 50,000 units of *L*-penicillin and 100 mg streptomycin should be added to each liter of medium. A few drops of P-2000 polypropylene glycerol should be added to prevent excessive foaming. Sterile conditions are not necessary.

5. Preparation of stock suspension

Starved cells, if kept under refrigeration in a well stoppered container, are usable for 4 — 5 days. Fresh suspensions for experimental use should be made up each day. These are made up in 0.1 M tris buffer, pH 7, and should contain about 10 mg (wet weight) of yeast/ml. The exact weight should be known.

II. Experimental Procedure

The experiment is divided into three parts. Students should work in pairs, each pair carrying out one part. The three parts of the experiment are as follows:

Part I: The effect of Rb^+ (2 mM) on the uptake of Cs^+ at different Cs^+ concentrations (1.25 — 20 mM).

Part II: The effect of Cs⁺ (1 mM) on the uptake of Rb⁺ at different Rb⁺ concentrations (low range − 0.4 to 6 mM).

Part III: The effect of Cs⁺ (1 mM) on the uptake of Rb⁺ at different Rb⁺ concentrations (high range − 8 to 50 mM).

All experiments are carried out at pH 7 (tris buffer) and room temperature. In each part, two sets of six tubes are set up. One set (1A − 6A)

Table 1

Tube	Part I			Part II		
	Cs*Cl (50 mM) groups A and B	Tris buffer group A	group B	Rb*Cl (10 mM) groups A and B	Tris buffer group A	group B
1	0.25	6.75	5.75	0.4	6.6	5.6
2	0.5	6.5	5.5	0.8	6.2	5.2
3	1.0	6.0	5.0	1.2	5.8	4.8
4	2.0	5.0	4.0	2.0	5.0	4.0
5	3.0	4.0	3.0	3.0	4.0	3.0
6	4.0	3.0	2.0	6.0	1.0	—

Part III		
Rb*Cl (100 mM) groups A and B	group A	group B
0.8	6.2	5.2
1.6	5.4	4.4
2.4	4.6	3.6
3.2	3.8	2.8
4.0	3.0	2.0
5.0	2.0	1.0

All quantities in ml. Reagents previously prepared and stored under refrigeration should be allowed to warm up to room temperature before beginning the experiment.

is a control. The second set (1B − 6B) contains the inhibiting cation. In each set, 2 ml of the stock yeast suspension containing 10 mg/ml and 1 ml of 1 M glucose are added to all tubes. (*Note*: the yeast suspension should be well shaken before removing each 2 ml aliquot). In addition, 1 ml of non-radioactive RbCl (20 mM) in tris buffer is added to each tube in group B of Part I, and 1 ml of non-radioactive CsCl (10 mM) in tris buffer is added to each tube in groups B of parts II and III. The additions listed in Table 1 are also made to make a final volume of 10 ml and a final yeast concentration of approximately 2 mg/ml in all tubes.

The experimental procedure in each part is as follows: First set up group A of the appropriate part of the experiment adding everything *except glucose*. Then add glucose to the first tube and begin timing the run from this point. 15 min after the addition of glucose filter out the cells by sucking the suspension through a millipore filter, previously moistened by sucking a few ml of ice cold distilled water through it. Using a plastic squeeze bottle, wash the cells twice under suction with ice cold water (use about 10 ml water for each washing). Direct the stream from the wash bottle to wash cells adhering to the sides of the filter holder down on to the filter paper but take care not to allow it to play directly on to the pad of cells already on the paper. When all excess water has been sucked through, carefully remove the millipore filter with its adhering pad of yeast cells from the filter holder and transfer it to a planchette containing 2 — 3 drops of 20% sucrose solution, making sure that the filter is firmly and evenly seated on the bottom of the planchette. Using an eye dropper, carefully place a drop or two of 5% glycerol on the cells. Place the planchette in the sample changer and count for ^{86}Rb or ^{137}Cs using appropriate settings of the pulse height analyzer [16].

Alternatively, the millipore filter with its adherent pad of cells can be removed from the filter holder with a long forceps, carefully introduced into a counting tube for an automatic well scintillation spectrometer and counted in this way.

Each part of the experiment should be run in duplicate so that each pair of students carries out 24 determinations of cation uptake. During the 15 min between adding glucose to tube 1A and filtering the contents of this tube, tubes 2A and 3A should be started at 5 min intervals. By continuing in this manner successive tubes can be started at intervals of 5 min throughout groups A and B of each experiment run. As soon as each sample is ready for counting it can be added to a sample holder in the counting apparatus. To avoid confusion, a distinctive marking scheme for the samples belonging to each pair of students in the class should be devised. It is convenient to have one student from each pair handle filtration, washing etc., his partner being responsible for adding glucose to each tube, timing the fermentations etc.

Appropriate blanks and standards should be made and counted for each part of the experiments. Blanks for background determination can consist simply of empty planchettes or well-scintillation sample tubes. Standards are prepared as follows: For planchette counting, place a millipore filter in a planchette containing 2 — 3 drops of 20% sucrose. Pipette in 1 ml of a 1/50 dilution of the appropriate radioactive stock solution and evaporate under a heating lamp until almost dry. Add

1 — 2 drops of 5% glycerol. Alternatively pipette 1 ml of the diluted radioactive solution into a well-scintillation counter sample tube. Standards may be prepared by the class instructor when the radioactive stock solutions are made. Duplicate or triplicate standards should be made for each part of the experiment.

III. Evaluation

Correct all counts for background. Using the corrected counts for the experimental samples and the mean corrected counts for the standards, compute the total uptake during the experiment (in μmol/g wet weight yeast) and the rate of uptake, V, (in μmol/g wet weight/hr) for Rb^+ or Cs^+ in each sample. From the total uptake and the known initial external concentration, (S), of the transported ion in each sample, verify the fact that (S) remained virtually constant during the experimental run (this is a necessary condition for the type of kinetic analysis used herein). Compute $1/V$ and $1/(S)$ for each sample.

Part I. Plot $1/V$ against $1/(S)$ for groups A and B. From the plots obtained estimate V_m and K_m for Cs^+ and K_i for Rb^+. Also compute a, $1 - a$ and $\dfrac{a}{1-a}$ for group B and plot the results according to the method of HUNTER and DOWNS [Eqs. (14) and (15)]. From the plot obtained make a second estimate of K_m for Cs^+ and K_i for Rb^+.

Part II. Plot $1/V$ against $1/(S)$ for groups A and B. From the results obtained with group A estimate K_m and V_m for Rb^+. From group B estimate V'_m (the apparent V_m under these conditions) and β (V'_m/V_m) for the effect of Cs^+ on Rb^+ uptake.

Part III. Plot $1/V$ against $1/(S)$ for groups A and B. Estimate V_m for Rb^+ from the data obtained for both groups, and K_m for Rb^+ from the results obtained with group A.

The results obtained by one pair of students performing each part of the experiment may now be combined to obtain the following information:

1. Compare the values of K_m and V_m for Rb^+ obtained in parts II and III with each other. Also compare these values with the K_i for Rb^+ found in part I. From the K_m values obtained for Rb^+ and Cs^+ calculate the relative affinities of the transport site for each of these ions, assigning an arbitrary value of unity to its affinity for Cs^+.

2. Compute a, $1 - a$ and $\dfrac{a}{1-a}$ for the combined results for groups B of parts II and III. Plot the data obtained according to Eq. (14). From the slope of the ascending segment of the curve compute the relative affinities of the modifier site for Rb^+ and Cs^+ (assigning a value of unity to Cs^+). Note that the absolute values for the dissociation constants for

the interaction of Rb^+ and Cs^+ with the modifier site cannot be obtained from the data of the present experiment. For a further discussion of this point see Ref. [3].

Statistical treatment of Michaelis-Menten kinetics: In kinetic experiments of the kind outlined above, the results may be subject to considerable random variations and statistical analysis of the data may be desirable or necessary. A useful discussion of the application of statistics to Michaelis-Menten kinetics will be found in Ref. [17]. If a sufficient number of independent estimates are available, and if time permits, the pooled results obtained from group A of parts I and II in this experiment may, at the instructor's discretion, be used to illustrate the statistical evaluation of K_m and V_m.

Acknowledgments

Grateful acknowledgement is made to a number of the advanced graduate students at the Department of Physiology, Indiana University Medical Center, and especially to Mr. John F. White, who tested this experiment under classroom conditions. Their experiences contributed materially to the final design of the experiment as described herein.

This report is based in part on work done at the University of Rochester under Contract W-7401-Eng-49 with the U.S. Atomic Energy Commission.

References

1. Michaelis, L., u. M. L. Menten: Biochem. Z. **49**, 333 (1913).
2. Wilbrandt, W., and T. Rosenberg: Pharmacol. Rev. **13**, 109 (1961).
3. Webb, J. L.: Enzyme and metabolic inhibitors, Vol. 1 (Chapters 2 and 5). New York: Academic Press 1963.
4. Lineweaver, H., and D. Burk: J. Amer. Chem. Soc. **56**, 658 (1934).
5. Hunter, A., and C. E. Downs: J. biol. Chem. **157**, 427 (1945).
6. Conway, E. J., and W. McD. Armstrong: Biochem. J. **81**, 631 (1961).
7. — Symp. Soc. exp. Biol. **8**, 297 (1954).
8. Rothstein, A.: In: A. M. Shanes, Ed., Electrolytes in biological systems, p. 65. Washington, D.C.: American Physiological Society 1955.
9. —, and M. Bruce: J. cell. comp. Physiol. **51**, 145 (1958).
10. Conway, E. J., and F. Duggan: Biochem. J. **69**, 265 (1958).
11. Rothstein, A., and M. Bruce: J. cell. comp. Physiol. **51**, 439 (1958).
12. Armstrong, W. McD., and A. Rothstein: J. gen. Physiol. **48**, 61 (1964).
13. — — J. gen. Physiol. **50**, 967 (1967).
14. Conway, E. J., P. F. Duggan, and R. P. Kernan: Proc. Roy. Irish Acad. **63**, B 93 (1963).
15. Monod, J., J. Changeux, and F. Jacob: J. molec. Biol. **6**, 306 (1963).
16. Oberhausen, E., and H. Muth: Double tracer techniques. (This volume.)
17. Wilkinson, G. N.: Biochem. J. **80**, 324 (1961).

Bacterial Permeases

By A. Kepes

According to Cohen and Monod (1956) the term 'permeases' designates specific transport mechanisms which are distinct from metabolic processes. Permeases usually participate in active transport: inside the cells the transported substrates are at a higher electrochemical potential than in the medium. The intracellular accumulation is associated with an increase of respiration and is sensitive to metabolic inhibition. When the rate of metabolism of a substrate exceeds that of its transport its intracellular concentration remains lower than that of the medium; in such instances permeases mediate downhill transport. Such processes do not require the provision of metabolic energy.

Permeases are characterized by a high degree of substrate specificity; they are susceptible to the action of poisons, and exhibit saturation kinetics. Bacterial cells are capable of synthesizing permeases for many different substrates but they actually synthesize only those for which substrates are available in the respective growth media. This 'induction' is subject to genetic control. Separate transport systems with individual permeases are assumed to exist when the following properties can be demonstrated:

1. Substrate Specificity. The transport of a given substrate is not competitively inhibited by known substrates of other permeases. Competing substrates are believed to be transported by the same permease.

2. Genetic Specificity. A single mutation leads to the loss of the capacity to transport all substrates of a given permease without affecting any other permease or the metabolism of the cell.

3. Specificity of Induction. A permease is induced only by one of its substrates but not by a substrate of a different permease. The experiments described below were designed to illustrate the methods which are commonly employed in studying specificity, kinetics, induction, and genetic control of bacterial permeases.

The technical problem consists in following the kinetics of β-galactoside permease mediated transport reactions (temperature dependence, Michaelis constant, action of inhibitors) in *E. coli* using methyl-thiogalactoside (TMG) or orthonitrophenyl galactoside (ONPG) as substrates and in demonstrating kinetic and environmental control of β-galactosidase and permease induction. In addition, accumulation and countertransport of α-methyl glucoside and isoleucine in *E. coli* is to be measured.

Outline of Experimental Procedure. The experiments described below will be divided between five groups of students, four of which will work on β-galactoside permease and the fifth on isoleucine and α-methylglucoside permease. Three different strains of *E. coli* will be supplied (Table 1):

a) one galactosidase+ − permease− inducible strain,

b) one β-galactosidase+ − permease+ inducible strain (wild type),

c) one β-galactosidase+ − permease+ constitutive strain.

β-galactoside permease activity will be estimated by two different techniques:

1. Measurement of accumulation of radioactive TMG by a rapid filtration technique.

2. Determination of the rate of hydrolytic conversion of the chromogenic substrate ONPG into a colored product by the intracellular galactosidase. Hydrolysis is a measure of permease activity since it proceeds potentially much faster than penetration.

The filtration technique will be employed also for the study of isoleucine and α-methyl glucoside transport.

Kinetic measurements are performed after arresting the growth of the bacteria by the addition of chloramphenicol.

Equipment

1. Special Equipment.

5 milipore filter holders on vacuum flasks connected to water jet pumps via three way stopcocks.

2. Standard Equipment.

1 spectrophotometer with monochromator at 420 mμ and 600 mμ.

1 scaler with automatic sample changer for 25 mm diameter C^{14} samples (thin window G.M. tubes) or liquid scintillation counter.

3 constant temperature baths each with shakers for a maximum of 14 erlenmeyer flasks of 250 ml or 100 ml capacity.

1 constant temperature bath for test tube racks (capacity 48 test tubes 18 × 180 mm and/or 22 × 220 mm).

5 stop watches.

Test tube racks.

5 ice boxes to keep one 1/2 l recipient in cold, millipore filters, 25 mm diameter, 0.45 μ pore size, code H.A.

Planchets for counting or liquid scintillation vials scintillating mixtures.
Warm room with shaker for 1 l erlenmeyer flasks for overnight cultures.

3. Glassware.

Erlenmeyer flasks: 1 l: 3; 500 ml: 1; 200 ml: 5; 100 ml: 32.
Pipettes: 20 ml: 5; 10 ml: 5; (blow out) 5 ml: 15; 2 ml: 3; 1 ml: 25; 0.5 ml: 10; 0.2 ml: 7; 0.1 ml: 18.
Test tubes: 18 × 180:30; 22 × 220:48.

4. Solutions.

IPTG (isotropyl-thio-galactoside)
TMG (methyl-thio-galactoside)
TDG (-D-galactoside -D-thio galactoside)
ONPG (orthonitrophenyl galactoside: 4 gr/l in 0.2 M/l sodium phosphate buffer pH 7)
Chloramphenicol 2.5 mg/ml
β-marcaptoethanol
Toluene
Deoxycholate: 1 gr/100 ml
Na_2CO_3 1 M/l
Lactose 0.1 M/l
PCMB 10^{-2} M/l (para hydroxy mercuribenzoate or para hydroxy mercuri-phenyl sulfonate or N ethyl maleimide)
NaN_3 1 M/l
L-isoleucine 10^{-2} M/l
D-isoleucine 10^{-2} M/l $\qquad$ α-methyl-glucoside 10^{-1} M/l
Glucose 1 M/l.

5. Isotopes.

C^{14} TMG
C^{14} α-methyl-glucoside
C^{14} isoleucine 10 µc.

6. Biological Material.

Bacterial strains.
Sterile culture medium: KH_2PO_4: 13.60 g; $(NH_4)_2SO_4$: 2 g; $MgSO_4$, 7 H_2O: 0.2 g; $FeSO_4$, 7 H_2O: 0.0005 g; glycerol: 4 g; adjust with KOH to pH 7.00; fill up with distilled water to 1.000 ml.
Sterilized in lots of 250 ml (3 − 4 l per session).
Sterile aqueous 20% glycerol in 10 ml tubes.
Sterile solution of vitamin B_1 5 mg/ml in 10 ml tubes.

4*

Execution

1. Preparation of Bacterial Cultures. The bacteria are grown aerobically on a simple mineral medium supplemented with glycerol. The growth of the three strains (cf. Table 1) is followed by measuring optical density at 600 mμ in a cuvette of 10 mm thickness. If plotted on semilog paper against time, the growth curves should be straight lines.

When the optical density of the cultures has reached 0.3 − 0.4 each of the cultures of ML 3 and ML 30 is to be subdivided into two portions, one of which is then supplemented with $2 \cdot 10^{-4}$ M isopropyl-thio-galactoside (IPTG), an inducer. At the time of addition of the inducer, a clock is started. 15, 40, and 90 min later samples of 10 ml each are withdrawn from each of the two flasks containing the inducer and further protein synthesis is arrested by the addition of a small volume of chloramphenicol

Table 1

Bacterial strains		Genotype	Phenotype				
			without inducer		with inducer		
ML	K 12		Z	Y	Z	Y	
ML 3	300	$i^+z^+y^-$	−	−	+	−	cryptic
ML 30	3000	$i^+z^+y^+$	−	−	+	+	wild
ML 308	3300	$i^-z^+z^+$	+	+	+	+	constitutive

to give a final concentration in the medium of 50 γ/ml. 10 ml samples are also withdrawn from the uninduced portions of the cultures of all three strains at a time when their optical densities reach about 0.8. After addition of chloramphenicol they are ready for use. In addition a minimum of 120 ml of the chloramphenicol arrested constitutive strain has to be prepared for use by groups 3, 4, and 5. The samples of chloramphenicol arrested cultures will be designated by two digit numbers, the first for the strain (1, 2, 3), the second for the induction time (0, 1, 2, 3). Samples 10, 13, 20, 21, 22, 23, and 30 provide cultures for groups 1 and 2. The other three groups share a minimum of 120 ml of the chloramphenicol arrested constitutive strain.

2. Assay Procedures. Throughout the experiments the bacteria have to be maintained at a constant temperature of 25 °C in the presence of a carbon source (e.g. glycerol). Vigorous agitation is necessary to ensure adequate O_2-supply.

In order to follow the time course of TMG or ONPG transport measurements of ^{14}C TMG incorporation and color development by ONPG hydrolysis respectively have to be made.

a) Filtration technique for measuring ^{14}C uptake.

In accurately measured time intervals after the addition of the substrate, TMG, (= zero time) 1 ml samples are withdrawn, if possible without interruption of shaking. The contents of the pipette are transferred into ice cold medium previously placed on the filter holder. The instant of emptying the pipette is the sampling time. The sample should be pipetted out of the culture flask not earlier than 15 sec before sampling time. Since the success of the experiment largely depends on the accuracy of sampling one has to go through the sampling procedure before the experimental runs. The following items must be ready for use:

Millipore filters, pincers for the manipulation of filter membrane, the various parts of the filter holder, precooled medium with a large mouthed pipette, clock, threeway stopcock to connect the filter holder to vacuum or atmosphere (the connections established by the various positions of the stopcock have to be recognized before the start of the experiment!), pipettes and flasks containing the required reagents, a suitable tray for deposition of the millipore filter after filtration. The places for the filters must be carefully labeled to ensure identification.

The time schedules given in the tables are designed for beginners, 1 min 30 sec are allowed for a cycle of millipore filtration including the positioning of filter membrane on filter holder, deposition of cold rince solution, sampling, establishment of vacuum, rincing twice, releasing vacuum removal of filter membrane from the filter holder to the tray. A first trial of these manipulations should be made without a bacterial sample in order to ascertain that the cycle can be achieved in 1 min 30 sec. For a trained experimenter the time can be decreased to 60 then to 40 sec. Special procedures have been worked out for faster sampling.

The timetables should be copied in large types for each experimenter and put in a position where a fast look allows to follow the programme.

Counting. Depending on the available equipment, counting can be made by way of a) a thin window G.M. tube or b) a liquid scintillation counter. For thin window counting the geometry of the sample changer dictates the choice of mounting.

Liquid samples (0.05 ml) for total counts are plated on planchets with the same geometry.

For scintillation counting the thoroughly dried millipore filters are transferred into scintillation vials. 10 ml scintillation liquid (toluene: 2000 ml; PPO: 4 g; POPOP: 0.2 g) is poured on the filter which becomes completely transparent.

Liquid samples for total counts are pipetted into the scintillation vials, and 10 ml of the following scintillation mixture is added (BRAY):

Dioxan:	880 ml
Naphthalene:	60 g
PPO:	4 g
POPOP:	0.2 g
Ethylene glycol:	20 ml
Methanol:	100 ml.

This mixture can incorporate up to 10% volume of aqueous solution.

b) Photometric determination of ortho-nitrophenyl galactoside hydrolysis: After addition of substrate (0.5 ml M/75 ONPG) to the buffered cell suspension (2.0 ml) = zero time, the development of color is observed. When the optical density approaches 1.0 the reaction is stopped by adding 1 ml of 1.0 molar Na_2CO_3. Note time. Read optical density at 420 mμ. The rate of hydrolysis is calculated by dividing the amount of dye released by the time elapsed until the termination of the reaction by Na_2CO_3.

3. Description of Experiments.

Group I. Genetic and physiological control of β-galactoside permease synthesis as tested by the filtration technique.

By the procedure described above (p. 52) cultures are obtained which were exposed for various lengths of time to the action of an inducer of β-galactoside permease, IPTG. Table 1 summarizes the properties of the three bacterial strains used. Samples of ML 30 (i.e. samples No. 20, 21, 22, 23) should exhibit increasing permease activity with increasing length of the induction period. In ML 3 (i.e. samples No. 10, 11, 12, 13) no permease activity should be inducible, whereas in ML 308 (i.e. samples No. 30, 31, 32, 33) a maximum amount of permease should always be present which cannot be affected by incubation in the presence of the inducer. These predictions are verified by the following experiments with samples No. 10, 13, 20, 21, 22, 23, and 30:

The seven cultures are shaken in a water bath at 25 °C. At zero time from each culture 5.0 ml is transferred by means of a blowout pipette into an Erlenmeyer flask containing 0.25 ml $2 \cdot 10^{-2}$ M ^{14}C TMG solution. Without interruption of shaking, 1 ml samples are withdrawn and filtered after incubation at 25 °C for 30 sec, 11 min, and 22 min. An example of a schedule for the seven parallel experimental runs is given in Table 2. After filtration each sample should be rinsed with about 10 ml of ice cold growth medium. The filters are dried and the amount of ^{14}C in the deposit is determined as described on p. 53. It is possible to convert the cpm's into concentration if 0.05 ml suspension from each flask is counted without previous filtration. The counts per min should be plotted against the time elapsed from the moment of addition of TMG. In addition the initial velocity i.e. the ^{14}C uptake within the first 30 sec of the experiment and the plateau values (average of the ^{14}C contents of the samples taken after 11 and 22 min) are plotted against the optical densities of the cultures as measured after stopping growth with chloramphenicol. The origin of these curves lies at the origin of the coordinate system since in the absence of any bacteria no TMG can be accumulated. The resulting curves should demonstrate that after addition of the inducer the increment of permease activity in strain ML 30 is proportional to the increase of the bacterial mass whereas in the constitutive strain (ML 308) permease is directly proportional to the actual bacterial mass.

The ratio of the increase of activity and the increase of bacterial mass is called the differential rate of synthesis. Strictly speaking, this rate should be expressed in mg of a given enzyme synthesized per mg of total protein synthesized in the same time. Our simplified evaluation of the data is based on the assumption that the measured activity is propor-

Table 2

Flask	10	13	20	21	22	23	30
Start	0	1'30	3'	4'30	6'	7'30	9'
Filter at	0'30	2'	3'30	5'	6'30	8'	9'30
	11'	12'30	14'	15'30	17'	18'30	20'
	22'	23'30	25'	26'30	28'	29'30	31'

tional to the mass of the enzyme and the optical density is proportional to the mass of total bacterial protein.

Group II. Kinetics and genetic control of galactoside permease activity as measured by following the downhill movements of ONPG by photometric determination of ONPG hydrolysis.

The experiments are performed with culture samples number 10, 11, 12, 13, 20, 21, 22, 23, and 30.

a) Genetics and regulation of β-galactosidase (Z)synthesis. Measurements of β-galactosidase activity in various strains grown with and without inducer. 0.05 ml of each of the cell suspensions is diluted by 2.0 ml of sodium phosphate buffer pH 7 containing β-mercapto-ethanol. β-galactosidase activity can be measured after destruction of the permeability barrier of the cells by toluenizing the cell suspensions: One drop of toluene and one drop of 1% deoxycholate is added and the suspensions are shaken for 20 min. After this treatment the enzyme is easily accessible for its substrate, ONPG. 0.5 ml of this substrate (M/75) is added to the medium (= zero time) and the resulting suspension is vigorously shaken in a water bath at 25 °C. When the optical density of tube 30 approaches 1.0 the reactions in all tubes are stopped by the addition of

1 ml molar Na_2CO_3. The reaction time is measured and photometric readings are taken at 420 mμ.

b) Genetics and regulation of permease (Y) synthesis. Measurements of permease activity in different strains grown with and without inducer. When the bacterial membranes are intact, the rate of ONPG hydrolysis is limited by the rate of uptake (i.e. by the activity of the permease). The effect of the permease is demonstrated by comparing the rates of ONPG hydrolysis in intact and toluenized cells. The uptake takes place mainly via the permease but passive leakage cannot entirely be neglected. In order to estimate the contributions of leakage to the results of this experiment, experiments of β-galactoside permease inhibitors are included and described under 'c'.

In order to estimate ONPG uptake in intact cells 0.4 ml of the above mentioned untreated cultures is added to each of two series of test tubes. 1.6 ml of mineral medium (without β-mercapto-ethanol) is added and one drop of 0.1 M TDG is added to one series. The reaction is started by the addition of 0.5 ml ONPG. Color development is watched in tube 30; the reaction is stopped, and time and optical density are estimated as described above.

c) Inhibition of permease. Prepare five tubes containing 0.5 ml ONPG and the following additions in a total volume of 2.1 ml made up with mineral medium (without β-mercapto-ethanol)

1. control (no additions);
2. thio-digalactoside (TDG) 10^{-3} M;
3. lactose 10^{-2} M;
4. para-hydroxymercuribenzoate 10^{-4} M;
5. sodium azide $2 \cdot 10^{-2}$ M.

Start reaction by adding 0.4 ml of culture sample 30, shake in thermostat. Stop reaction with carbonate. Read at 420 mμ. Note time of start and stop.

d) Km of permease for ONPG. Prepare two sets of test tubes containing the following volumes of M/75 ONPG: 0.05, 0.1, 0.2, 0.3, 0.4, 0.5 ml. Make up volume with mineral medium to 2.1 ml. To one set of tubes add one drop of inhibitor (TDG 10^{-1} M or PCMB 10^{-2} M). Start reaction by adding to each tube 0.4 ml of culture 30. If possible assays *c* and *d* should be started and stopped at the same time.

Evaluation of data: For series *a* and *b* plot rate of ONPG hydrolysis versus optical density of the culture at the time of addition of chloramphenicol (differential rate of synthesis of *Z* and *Y* respectively, see remarks group I above).

For series *d* plot rate of hydrolysis after substraction of the background (inhibitor present) versus ONPG concentration in direct and double reciprocal scale (Lineweaver Burk plot).

Group III. TMG transport: time course, inhibitors, and Michaelis constant. For this and all subsequent experiments use culture 30 (strain ML 308) at optical density 0.800 (c.f.p. 55) and a stock solution of $4 \cdot 10^{-2}$ M ^{14}C-TMG containing 0.1 mCi/millimole.

a) Time course and inhibitors. In a thermostatic bath adjusted to 25 °C prepare flask A (capacity 250 ml) with 0.62 ml of stock solution of TMG, and flask B, C, D of 100 ml capacity each with respectively 0.1 ml of TDG, 0.1 M, PCMB, 0.01 M. NaN$_3$, 1.0 M. At zero time pipette into flask A 25 ml of culture. After 12, 14, and 16 min 5 ml aliquots of this culture will be transferred to flasks B, C, and D respectively. 1.0 ml samples are filtered according to the following schedule (Table 3) one experimenter samples from A another from B, C and D:

Table 3

	A 25 ml from culture	B 5 ml from A	C 5 ml from A	D 5 ml from A
Start	0′	12′	14′	16′
Filter 1 ml at	0′30	12′30	14′30	16′30
	2′	18′	19′30	21′
	4′	22′30	24′	25′30
	8′			
	10′			
	12′30			
	14′30			
	17′			
	20′			

The resulting curves show the time course of TMG uptake (curve A) and the effects of the various poisons on various stages of the accumulation process (curves B, C, D).

b) Km of permease for TMG. TMG uptake is measured in 100 ml Erlenmeyer flasks at 25 °C in the presence of the following concentrations of TMG in the incubation media.

$$A, \; 2 \cdot 10^{-3},$$
$$B, \; 4 \cdot 10^{-4},$$
$$C, \; 2 \cdot 10^{-4},$$
$$D, \; 8 \cdot 10^{-5},$$
$$E, \; 4 \cdot 10^{-5} \; \text{mol/l.}$$

These concentrations are obtained if

TMG stock solution A 0.25 ml,
                      B 0.05 ml,

$$C \quad 0.25 \text{ ml,}$$

and ten fold diluted stock solutions $D \quad 0.1$ ml,

$$E \quad 0.05 \text{ ml,}$$

are each mixed with 5 ml culture. The experiment is started by the addition of the culture to the TMG solutions.

Schedule (Table 4):

Table 4

	A	B	C	D	E
Start by adding 5 ml culture at	0′	2′	4′	6′	8′
Filter at	0′30	2′30	4′30	6′30	8′30
	10′	11′30	13′	14′30	16′
	17′30	19′	20′30	22′	23′30

In addition, from each flask 0.05 ml aliquots are withdrawn for counting. These 'total counts' yield the TMG concentration of the incubation media. Plot ^{14}C uptake as measured after 30 sec and plateau values (i.e. the average of the uptake observed in the other samples) versus total counts on direct and double reciprocal scale.

Group IV. Temperature dependence of β-galactoside permease. Two 100 ml flasks containing 0.5 ml stock solution of TMG are prepared in constant temperature bath at 15° and 37° respectively. In addition 12 to 15 ml aliquots of the culture are equilibrated on the respective temperatures.

Table 5

	15°	37°
Start	0′	0′
Filter 1 ml at	1′	0′30
	3′	2′
	6′	3′30
	9′	5′
	12′	6′30
	15′	8′
	20′	10′
	25′	15′
	30′	20′

To each flask is added 10 ml culture at zero time and 1 ml aliquots are filtered on millipore according to the time schedule of Table 5. The two experiments are independent in timing and could be done one after another, or at the same time by two different operators.

Results should be plotted together with curve A group III at 25°.
Group V. Accumulation and counter flow of α-methyl glucoside and isoleucine. The cells are first allowed to accumulate either ^{14}C α-methyl glucoside or ^{14}C isoleucine until equilibrium is established. Subsequently non radioactive substrates such as α-methyl-glucoside or glucose, and L or D isoleucine are added and the redistribution of radioactivity is followed (for theoretical treatment of counterflow cf. WILBRANDT, chapter 4). The experiment is performed at 25 °C. Six 100 ml flasks will be prepared. Exp. A, B, C is independent from exp. D, E, F. They contain:

A 0.45 ml of 0.01 M/l ^{14}C α-methyl-glucoside;
B 0.10 ml of 0.1 M/l non radioactive α-methyl-glucoside;
C 0.10 ml of 0.01 M/l glucose;
D 0.15 ml of 0.01 M/l ^{14}C isoleucine (racemic);
E 0.10 ml of 0.1 M/l non radioactive L-isoleucine;
F 0.10 ml of 0.1 M/l non radioactive D-isoleucine.

First, ^{14}C uptake is followed in flasks A and D. Uptake is started by the addition of 15 ml of culture to each flask (optical density 0.8, any strain will do since they differ only with respect to lactose metabolism). After about 12 — 15 min of ^{14}C uptake, 4 ml aliquots are transferred from A to B and C and from D to E and F. This initiates counterflow. Sampling is done by filtering 1 ml aliquots according to the following schedule (Table 6):

Table 6

	A	B	C	D	E	F
	Add 15 ml culture	Add 4 ml from A	Add 4 ml from A	Add 15 ml culture	Add 4 ml from D	Add 4 ml from D
Start	0'	12'	14'	0'	12'	14'
	0'30	12'30	14'30	0'30	12'30	14'30
	2'	17'30	19'	2'	17'30	19'
	4'	21'	23'	4'	21'	23'
	7'			7'		
	10'			10'		
	16'			16'		

0.05 ml are withdrawn from A and D for the estimate of total counts. Plot cpm's incorporated into the cells versus time.

Comments

So far the results of all ^{14}C estimates have been expressed in counts per min recovered in the cells of the filtered samples. It would be preferable however to give the results in μmoles per g dry weight of cells or in terms of intracellular concentrations. The conversion of the data into these units can be accomplished as follows:

The optical density of bacterial suspensions is very nearly proportional to dry weight. It varies little with size and number of cells. An optical density of 1.00 at 600 mμ (1 cm light path) was found to correspond to 275 μg dry weight per ml of suspension. The conversion factor depends on the type of photometer, the wavelength and the bacterial species.

The specific activity of the isotopically labelled solutions is deduced from the 'total' counts at known concentration in every experiment.

Specific activity in cpm per μmol

$$= \frac{\text{cpm (total counts)}}{\text{Volume of sample in ml} \times \text{μM/ml in experimental mixture}}$$

Intracellular concentration in μmoles/g dry cells

$$= \frac{\text{Counts per min on millipore}}{\text{Volume of sample filtered} \times \text{specific activity} \times \text{dry weight per ml}}$$

Concentration in μMol per ml cell water would be about one fourth of this value. For an experiment where the final TMG concentration amounts to 10^{-3} M, where 1 ml samples are filtered, and where 0.05 ml samples were taken for total counts, the intracellular concentration amounts to

$$\frac{\text{cpm (millipore)}}{\text{cpm (total counts)} \times 20 \times \text{opt. density} \times 275 \times 10^{-6}} \text{ μmoles/g}.$$

It is convenient to plot first time course of uptake in terms of cpm versus time, to convert the data into μmol/g dry weight subsequently, and then to change the scale of ordinates to μmol/g without making a new graph.

The experiments showing genetic control of inducibility of galactosidase and permease in the various strains of E. coli are self explanatory and need no further comments. The same applies to the demonstration of β-galactosidase activity inside the cells.

Three measures of permease activity were used.

1. The initial velocity of TMG uptake,

2. the steady state value of intracellular TMG reached at the end of accumulation process. This method is based on the observation that the size of the intracellular steady state pool is closely proportional to the rates of uptake as measured by method (1). Although this method seems theoretically objectionable it gives far more accurate results than method (1).

3. The in vitro hydrolysis of ONPG.

In strain 300 P, or ML 3, a permease-less point mutant, this method often shows a non negligible activity although strictly negative results are always obtained with TMG.

Three kinds of inhibitors are tested with both methods. A competitive inhibitor TDG with the proper β-galactoside configuration is itself transported actively as can be shown by independent experiment. An SH group reagent inactivates the permease protein. This can be partially reversed by an excess of thiol compounds. Independent experiments show that the SH group which is essential for permease activity, can be protected by the presence of excess substrate (TDG). This served for the differential labeling and isolation of the permease protein by Fox and KENNEDY (1965).

The third inhibitor, sodium azide is a poison of energy metabolism. It inhibits active accumulation of TMG but hardly downhill transport of ONPG. This shows that the energy coupling step is not essential for the transfer through the membrane, although of course accumulation of TMG requires continuous energy expenditure which goes on even in the steady state when active inward transport is balanced by passive outflow. The applicability of simple Michaelis kinetics suggests that in the concentration ranges explored, one and the same permease-substrate complex is the rate limiting intermediate for transport. This is true also for different degrees of induction of permease, since progress of the induction process does not cause any detectable change of K_m.

Virtually the same K_m's are found for the initial velocity of TMG uptake and the plateau value of accumulation. This is a most stringent quantitative indication of the fact that the size of the steady state pool is proportional to the velocity of uptake. This implies that the outflow from the cells occurs at a rate proportional to the pool concentration. This conclusion is further supported by the finding that the exit process is a first order reaction.

The proportionality between rate of uptake and final pool concentration does not hold at all temperatures. At lower temperatures the rate of uptake decreases whereas the plateau value concomitantly increases. Thus the exit process has a higher temperature coefficient than uptake. Besides the demonstration of counterflow this is the best evidence for the view that exit does not take place by free diffusion but by carrier mediated passive transport.

Counterflow of TMG can be provoked by the homologous substrate β-galactoside as well as by a heterologous substrate, glucose. The effect of glucose can be observed even with galactoside permease negative strains; this is a strong indication that the carrier is independent of the permease and less specific.

Summarizing, the various results of the kinetic approach for elucidating permease mechanisms suggest that at least three kinds of molecules are involved in the transport cycle: a specific permease protein responsible for substrate recognition and selectivity; a less specific carrier which combines reversibly with the substrate to form a molecule which is physically capable of crossing the permeability barrier; and an energy coupling enzyme which enables the cells to perform transport against electrochemical gradients.

References

BRITTEN, R. J., and F. T. McCLURE: Bact. Rev. 26, 292 (1962).
COHEN, G. N., and J. MONOD: Bact. Rev. 21, 160 (1957).
ENGLESBERG, E.: Cold Spr. Harb. Symp. quant. Biol. 26, 261 (1961).
FOX, C. F., and E. P. KENNEDY: Proc. nat. Acad. Sci. (Wash.) 54, 891 (1965).
KEPES, A.: Biochim. biophys. Acta (Amst.) 40, 70 (1960).
— C. R. Acad. Sci. (Paris) 244, 1550 (1957).
—, and G. COHEN: In: GUNSALUS and STANIER, Ed., The bacteria, Vol. IV, p. 179. New York: Acad. Press 1962.
—, et J. MONOD: C. R. Acad. Sci. (Paris) 244, 809 (1957).
KOLBER, A. R., and W. D. STEIN: Nature (Lond.) 209, 691 (1966).
MITCHELL, P., and J. MOYLE: J. gen. Microbiol. 9, 257 (1953).
RICKENBERG, H. V., G. N. COHEN, G. BUTTIN et J. MONOD: Ann. Inst. Pasteur 91, 829 (1956).
SISTROM, W. R.: Biochim. biophys. Acta (Amst.) 29, 579 (1958).

The Exchange of ^{22}Na between Frog Sartorius Muscle and the Bathing Medium

By P. C. CALDWELL and R. D. KEYNES

I. Introduction

a) Description of Problem. To investigate the rate of exchange of labelled sodium between the extracellular compartment of frog muscle, the interior of the muscle fibres, and the surrounding medium.

b) Principle of Method. The two sartorius muscles are dissected from a frog and detached at the pelvis. They are then immersed in Ringer's solution containing radioactive sodium, one for 5 min and the other for several hours. In the muscle immersed for only a short period, the labelled sodium exchanges mainly with the sodium in the fibre interspaces, whereas in its companion there is time for most of the intracellular sodium to be exchanged as well. The time course of the two phases of sodium exchange is examined by washing out the radioactivity in a succession of samples of inactive Ringer's solution. The loss of ^{22}Na from the interspaces is complete after about 20 min, and the subsequent loss represents the efflux of intracellular sodium.

c) Further Variations in the Experiment. If the cardiac glycoside ouabain, which is a specific inhibitor of the sodium pump, is added to the Ringer's solution as soon as the magnitude of the resting efflux has been established, its action on the sodium efflux can be seen. The effect on the efflux of removing external sodium can also be examined, by substituting lithium for all the sodium in the Ringer's solution.

II. Instruments and Solutions

Dissecting microscope.

Scintillation counter with well-type crystal detector for γ-radiation.

Small glass vessels to fit the counter.

Solutions:

1. Normal frog Ringer for dissection, with [K] = 2.5 mM (111 mM-NaCl, 1.8 mM-CaCl$_2$, 1.11 mM-K$_2$HPO$_4$, 0.28 mM-KH$_2$PO$_4$).

2. Normal frog Ringer containing ^{22}Na (specific activity of the order of 10 μc/ml).

3. High-K frog Ringer for the experiment, with $[K] = 10$ mM in order to increase the contribution of ouabain-sensitive active transport of sodium to the total sodium efflux (111 mM-NaCl, 1.8 mM-CaCl$_2$, 7.5 mM-KCl, 1.11 mM-K$_2$HPO$_4$, 0.28 mM-KH$_2$PO$_4$).

4. High-K frog Ringer as in (3) with the addition of 10^{-4} M ouabain.

5. High-K ouabain Ringer as in (4) made up with 111 mM-LiCl instead of 111 mM-NaCl.

III. Execution

Dissection. Free the muscles at the knee, and tie a cotton thread around the tendon, taking care not to crush any of the muscle fibres in the knot. Then free the muscles up to the attachment at the pelvis, cutting through with scissors the thin sheet of connective tissue on either side of the muscle, and the blood vessels which enter from below, rather than tearing them. Be prepared for a twitch on cutting through the nerve which enters the muscle about one third of its length from the knee. The tendon at the pelvic end is very short, and must be cut through as close as possible to the bone if the muscle fibres are not to be damaged in the process. In order to obtain easily observable changes in the efflux when ouabain is applied or lithium is substituted for sodium, it is advisable to use small frogs, e.g. *Rana temporaria* rather than *R. esculenta*. For best results, the muscles should be tied to thin glass or polyethylene rods so as to keep them stretched to their resting length in the body, that is to about 1.2 times their unstretched length. If this is attempted, it will be essential to use a microscope when tying the very short pelvic tendon of the muscle to the rod, and some practice will be needed in order to avoid drawing the ends of the muscle fibres into the knot and thus damaging them.

Loading the Muscles with ^{22}Na. Immerse the muscles in the radioactive Ringer at room temperature. Handle them by means of the threads or by the rods, and do not allow the far ends of the threads or rods to get into the radioactive solution. Leave one muscle in ^{22}Na Ringer for only 5 min, and then start on the washing-out procedure described below. Leave the other muscle in ^{22}Na Ringer for not less than an hour, or for the time taken to complete the first series of counts.

Collection and Determination of the ^{22}Na Washed out of the Muscles. Measure out $5 - 10$ ml of the high-K Ringer's solution into a number of the glass vessels. The exact volume chosen will depend on the capacity of the counter's well, and the pipetting must be accurate so that the samples will all be counted with equal efficiency. If the rod technique is used, the depth of the solution must be great enough to cover the whole of the muscle. Transfer the first muscle direct from the labelling solution into one of the vessels, allowing it to drain so that a minimum volume of the radioactive solution is transferred with it. After exactly 5 min, move the muscle on to vessel No. 2, then to No. 3, and so on. After six successive collecting periods of 5 min, increase the collecting time to 10 min for vessels Nos. 7 to 9. On completion of sample No. 9, put the muscle in a small beaker, add a few drops of concentrated nitric acid, and then digest the muscle by gentle heating for a short while. When digestion is complete, add sufficient Ringer's solution to make the total volume the same as that of the other samples, and count the digested muscle in

Table 1

Sample No.	Count	Counting time (sec)	c/sec	c/sec less background ($= 2.93$)	Total c/sec in muscle	c/sec $\cdot$ min	k_2 (min^{-1})
1	59811	100	598.1	595.2	1027.3	59.52	0.0579
2	20753	100	207.5	204.6	627.4	20.46	0.0326
3	10831	100	108.3	105.4	472.4	10.54	0.0223
4	7288	100	72.9	70.0	384.7	7.00	0.0182
5	5610	100	56.1	53.2	323.1	5.32	0.0165
6	4509	100	45.1	42.2	275.4	4.22	0.0153
7	4000	122.2	32.73	29.80	239.4	2.98	0.0124
8	4000	174.2	22.96	20.03	214.5	2.00	0.0093
9	4000	205.1	19.50	16.57	196.2	1.66	0.0085
10	4000	221.9	18.03	15.10	180.4	1.51	0.0084
11	4000	424.4	9.43	6.50	169.6	0.65	0.0038
12	4000	849.1	4.71	1.78	165.5	0.18	0.0011
13	4000	952.2	4.20	1.27	164.0	0.13	0.0008
14	4000	980.6	4.08	1.15	162.8	0.115	0.0007
15	4000	355.6	11.25	8.32	158.1	0.832	0.0053
16	4000	295.0	13.56	10.63	148.6	1.063	0.0072
17	4000	312.5	12.80	9.87	138.4	0.987	0.0071
Muscle	13642	100	136.4	133.5	133.5	—	—

order to determine the amount of radioactivity remaining in the muscle at the end of the washing-out period.

By the time these counts are done, the second muscle will have spent 1 — 2 h in the labelling solution, and will be ready for observation of the efflux of sodium from the intracellular compartment. In this case, it is convenient to use 10 min collecting periods throughout, changing the solutions according to the following schedule:

Samples 1 — 6   High-K Na Ringer,
samples 7 — 10 High-K Na Ringer + 10^{-4} M ouabain,
samples 11 — 14 High-K Li Ringer + ouabain,
samples 15 — 17 High-K Na Ringer + ouabain.

After completion of sample No. 17, digest the muscle in nitric acid and determine the radioactivity still remaining in it in the same way as before.

IV. Calculation of Results

The results can be treated in two different ways:

a) Plot of Counts Remaining in Muscle Against Time. If the amount of radioactivity left in the muscle at the end of the experiment is A counts/sec, and if the amount of radioactivity in the final Ringer sample is B_n counts/sec, then the activity in the muscle at the mid-time of the last (n-th) collecting period is given approximately by $A + \frac{1}{2} B_n$. At the mid-time of the penultimate, $(n-1)$th, collecting period, the total activity is

$$A + B_n + \frac{1}{2} B_{n-1}$$

and so on. It is thus a simple matter to work out backwards from the final sample the value of the total radioactivity in the muscle at the mid-time of each collecting period. When this has been done, the results can be plotted on semilogarithmic paper, with radioactivity on the logarithmic scale and time from the moment of removal from the radioactive solution on the linear scale.

For the outward movement of labelled sodium from a single compartment into an unlabelled solution we have

$$d[\text{Na*}]/dt = - k[\text{Na*}]$$

where t is time, and k is the rate constant for loss of Na* from the compartment. It follows that

$$[\text{Na*}] = [\text{Na*}]_0 \cdot e^{-kt}$$

or

$$\log_e[\text{Na*}] = \log_e[\text{Na*}]_0 - kt$$

where $[\text{Na*}]_0$ represents the amount of Na* in the compartment at zero time. For such a system, a plot of log radioactivity against time is a straight line whose slope gives the value of k, and whose intercept at zero time gives the value of $[\text{Na*}]_0$.

However, the individual muscle fibres are not, except at the outer surface of the muscle, in direct contact with the bathing solution, so that a whole muscle will not behave as a single-compartment system, but rather as an intracellular compartment in series with an extracellular compartment. The loss of Na* will be represented approximately by

$$[\text{Na*}] = [\text{Na*}]_1 \cdot e^{-k_1 t} + [\text{Na*}]_2 \cdot e^{-k_2 t}$$

where the suffixes 1 and 2 indicate the initial contents of Na* and the rate constants of the extra- and intracellular compartments respectively. Plotted semi-logarithmically, the radioactivity therefore declines along two straight lines in succession, and rough values for k_1 and k_2 can be obtained from the initial and final slopes, and for $[Na^*]_1$ and $[Na^*]_2$ from the intercepts at $t = 0$. Examine the results plotted in this fashion for the two muscles, and observe the effect of labelling the intracellular sodium more completely in one case than the other.

It must be stressed that this description has been oversimplified in several respects. The errors arising when $[Na^*]_1$ and $[Na^*]_2$ are thus determined by a crude extrapolation procedure may, as has been pointed out by HUXLEY (1960), be far from negligible, particularly if k_1 and k_2 are not very different. Other factors which should not be forgotten are the inhomogeneity of the muscle fibre population, and the possibility that the sodium classified as 'intracellular' is in effect contained in more than one compartment. Given that for a plane sheet of tissue of thickness $2b$, exposed to the medium on both sides, the rate constant for loss of the final two thirds of the extracellular radioactivity is given by

$$k_1 = 2.5 \, D'/b^2$$

work out the value of the effective diffusion constant for sodium in the muscle interspaces, D'. Determine b from the dimensions and wet weight of one of the muscles. How does the value of D' compare with the self-diffusion coefficient for Na^+ ions in free solution, which is about 10^{-5} cm^2/ sec at room temperature? For a discussion of some of these problems, refer to KEYNES (1954), and see also KEYNES and STEINHARDT (1968).

b) Plot of Rate Constants against Time. When the total radioactivity in the muscle is plotted semi-logarithmically against time, as in method (a) just described, changes in sodium efflux during the experiment appear as changes in slope, and unless they are large are often not very obvious. When examining the effects on the efflux of inhibitors, or of changes in the composition of the external medium, it is better to plot the rate constants directly against time on a linear scale. Since

$$k_2 = - \, (1/[Na^*]_2) \cdot d[Na^*]_2/dt$$

it follows that the rate constant for the n-th collecting period is B_n/T $(A + \frac{1}{2} B_n)$, where T is the duration of the collecting period. For the $(n - 1)$th period, the rate constant is $B_{n-1}/T(A + B_n + \frac{1}{2} B_{n-1})$ and so on. Thus it is simple, working backwards as before, to calculate the rate constant at the mid-time of each of the collecting periods. An example of the best way of laying out the experimental results in order to process them like this is given below. What can be concluded about the actions on the sodium efflux of (1) ouabain, and (2) removal of external sodium? What are the possible implications of the effect of the lithium solution? (See KEYNES and SWAN, 1959; KEYNES, 1965, 1966.)

5*

V. Recommended Method of Tabulating Results

The figures shown in Table 1 were obtained in an actual experiment on a muscle well loaded with ^{22}Na, and illustrate the most convenient way of handling the results. Conditions were optimal, and it is unlikely that ouabain and lithium will always give such large effects. Note that the counting time was adjusted so that at least 4000 counts were recorded for each sample, giving a standard deviation never greater than $\pm 1.6\%$. A somewhat lower accuracy — say not less than 1000 counts for each sample — would normally be acceptable. The collecting period was 10 min for every sample.

References

HUXLEY, A. F.: Appendix 2 to chapter on compartmental methods of kinetic analysis by A. K. SOLOMON. In: COMAR, C. L., and F. BRONNER, Eds., Mineral metabolism, Vol. 1A, pp. 163—166. New York: Academic Press 1960.

KEYNES, R. D.: The ionic fluxes in frog muscle. Proc. roy. Soc. B **142**, 359—382 (1954).

— Some further observations on the sodium efflux in frog muscle. J. Physiol. (Lond.) **178**, 305—325 (1965).

— Exchange diffusion of sodium in frog muscle. J. Physiol. (Lond.) **184**, 31—32P (1966).

—, and R. A. STEINHARDT: The components of the sodium efflux in frog muscle. J. Physiol. (Lond.) **198**, 581—599 (1968).

—, and R. C. SWAN: The effect of external sodium concentration on the sodium fluxes in frog skeletal muscle. J. Physiol. (Lond.) **147**, 591—625 (1959).

Determination of Intracellular Ionic Concentrations and Activities

By A. Kleinzeller, P. G. Kostyuk, A. Kotyk, and A. A. Lev

A knowledge of intracellular activities is the prerequisite for an appraisal of concentration and electrochemical gradients established in living systems by their physical properties and metabolic activities. While the determination of cellular ionic concentrations appears at first to be relatively simple, it is in fact complicated by a number of factors:

1. Structural compartmentation of cells and tissues: This term involves not only the extracellular space in the analyzed biological material, but also intracellular structural compartmentation (subcellular components with distinct permeability barriers) as well as a possible non-homogenity of the analyzed cell population.

2. Chemical compartmentation, affecting the ionic activity in a given physical compartment, such as 'binding' of ions or availability of water as a solvent. While several approaches may be used for the determination of intracellular ionic concentrations or activities, the values obtained are mostly only approximations, and the term 'apparent concentrations' or '-activities' appears to be justified.

As a first approximation, two structural compartments have to be considered in tissues and cell suspensions:

a) The extracellular space E: Under steady-state conditions in vitro, it may be safe to assume that the ionic concentrations and activities in the extracellular space are identical with those in the surrounding medium. The size of this space is taken to be identical with that occupied at equilibrium by water-soluble substances of molecular weight sufficient to prevent their entry into the cells.

b) The intracellular space: Assuming a uniform distribution of ions in intracellular water, the apparent intracellular ionic concentration may then be computed, based on an analysis of tissue water and the given ionic species, E, and the ionic concentration in the external medium.

The steady-state kinetics of labelled ionic species may provide information as to the number of diffusion barriers and the size of the corresponding spaces. Selective microelectrodes introduced into the cells may serve to measure intracellular activities.

The methods used for the determination of apparent intracellular ionic concentrations and activities will be discussed here; the kinetic approach will be dealt with in another course (c.f. Caldwell and Keynes, p. 63).

I. Determination of Apparant Intracellular Ionic Concentrations by Chemical Analysis

Problem. Measurements of intracellular potassium and sodium, and extracellular (Inulin) space in frog sartorius muscle.

Principle of Method. From the known concentrations of K, Na, and inulin in the incubation medium and the measured amounts of these substances in the tissue and the tissue water content, the amount of intracellular water and the intracellular ion concentrations are calculated.

Instruments. Torsion balance, analytical balance, flame photometer, centerwell scintillation counter, flame photometer, desiccator (with concentrated H_2SO_4), drying oven adjustable to 110 °C, ash free hardened filter paper (e.g. Schleicher and Schuell No. 589/3 or Whatman No. 541), boiling water bath, pyrex test tubes, 100 ml Erlenmeyer flasks, volumetric flasks (15 and 10 ml), 10 frogs.

Solutions. Concentrated HNO_3, 6N NaOH, Ringer's solution, ^{131}I inulin, NaCl and KCl standard solutions for flame photometry.

1. Determination of the Steady State Tissue H_2O and Bulk Cations (Na, K)

A steady-state distribution of tissue water and electrolytes is obtained by a sufficiently long incubation of the tissue in vitro in the required medium; the essential conditions of in vitro work (e.g. suitable choice of thickness of tissue, substrate, gaseous phase, etc., see Umbreit, Burris and Stauffer, 1957; Kleinzeller, 1965) have to be adhered to.

10 frog sartorius muscles are incubated at room temperature in a conical 100 ml flask containing 10 ml of frog ringer solution; gaseous phase: air; incubation in a Dubnoff metabolic incubator for 4 h.

After incubation, the tissue is removed from the medium, gently blotted for about 30 sec with a disk of ash-free hardened filter paper, trimmed to remove tendons, and weighed on a torsion balance. Five muscles are then transferred to preweighed vials and dried for 4 − 5 h at 110 °C; after cooling the vials in a desicator over H_2SO_4 for 30 min, the vials are weighed again. From the loss of weight during drying tissue water content is calculated, and expressed in kg/kg dry weight (DW)

$$\text{kg } H_2O_t/\text{kg } DW = \frac{\text{mg tissue wet weight}}{\text{mg tissue } DW} - 1.$$

A further five muscles are also blotted, trimmed and weighed, and are then transferred into pyrex test tubes. 0.2 ml concentrated HNO_3 is added to each tube and the tubes are placed in a boiling water bath for 15 min; in this way the tissue is ashed. The tube contents are diluted to a desired volume with redistilled water (for sartorii of about 50 mg weight a volume of 15 ml is usually convenient). In the clear extract Na and K are determined by flame photometry. The values obtained are expressed in mequiv./kg *DW*.

A sample of the incubation medium is also diluted and both cations are determined.

Comments

a) The above method of ashing the tissue cannot be used if tissue Cl is also to be determined. The following procedure is recommended: The dry tissue remaining in the vials after analysis of tissue water is quantitatively transferred into a pyrex test tube. The dry tissue is crushed with a glass rod and the electrolytes are extracted:

4 ml redistilled H_2O are added and the suspension is placed in a boiling water-bath for 15 min; after cooling, 2N H_2SO_4 is added to bring the final concentration to 0.2 N and the extract is diluted as required. The extraction of electrolytes is complete after 24 h. In the clear extract Na and K are determined by flame photometry, Cl by a potentiometric titration with $AgNO_3$ (see RAMSAY et al., 1955). In this way, water and tissue electrolytes can be determined in the same tissue sample.

b) Wet tissue may be directly extracted with dilute acid (0.2N HNO_3 or H_2SO_4) for at least 24 h; tissue water has to be determined in separate samples.

c) Cations may be determined by ashing the dry or wet tissue in a Pt crucible at 650 °C overnight; the ash is dissolved in 1 ml N HNO_3 and the appropriately diluted extract is used for the determination of Na, K, Ca and Mg. The alkaline earths are analyzed either by flame photometry or, better, by atomic adsorption spectroscopy or by chemical methods (see e.g. HÖFER, 1963).

2. Determination of the Extracellular Inulin Space

4 to 6 frog sartorii are incubated in approximately 10 ml of a frog Ringer solution containing inulin labelled with [131]I (approximately 1.0 μC/ml). The conditions of incubation are identical with those described for the determination of the steady-state ionic distribution. After incubation, the tissue is gently blotted on filter paper, rapidly weighed and transferred to 10 ml volumetric flasks containing 1.0 ml 6N NaOH. The flasks are kept in a boiling water bath for 3 min. The resulting turbid

colored solution is filled up to the mark with distilled water and a 1-ml-sample is pipetted into a 20-ml-plastic-vial and counted in the centerwell of a crystal probe for 2 min (Panel settings of the Packard autogamma system: gain: 10%, window: lower limit 50, upper limit 1000).

Simultaneously, the activity of a suitably diluted sample of the incubation medium is also determined.

From the data, the inulin space E in kg H_2O/kg tissue wet weight is calculated:

$$E = \frac{\text{counts/min in 1 g of tissue}}{\text{counts/min in 1 ml medium}}.$$

The extracellular fraction of tissue water H_2O_E is then:

$$E \cdot \frac{\text{tissue wet weight}}{\text{tissue water}} = E \cdot \frac{H_2O_t + 1}{H_2O_t}.$$

(H_2O_t denoting the value of tissue water in kg/kg DW, as determined from the difference of wet and dry weight), and the intracellular fraction of tissue water H_2O_I equals: $H_2O_I = 1 - H_2O_E$, both values in kg.

Comments

a) The extraction of inulin from the tissue for analysis may be carried out in a number of ways:

1. Wet tissue, blotted and weighed as described above, is transferred to a test tube containing 3 ml destilled water; inulin is quantitatively extracted from tissue slices within 24 h.

2. The extraction may be aided by homogenizing the tissue in a small Potter-Elvehjem apparatus in the presence of 2 ml 5% (w/v) trichloro-acetic acid: the suspension is left standing for 1 h and then centrifuged; the sediment is again extracted for 1 h with 2 ml 5% TCA, and the combined extracts are used for inulin determination.

Inulin may be determined chemically by any method specific for fructose; the methods of Cole (see Bacon and Bell, 1948) and of Kulka (1956) have proved satisfactory. A tissue blank has to be deducted from the values obtained. ^{14}C-inulin may be used and instead of the colorimetric determination, radioactivity measurements are carried out.

b) The inulin space is usually smaller than that determined for other 'non-permeable' substances, e.g. sucrose, sulphate, etc. In some tissues (e.g. kidney cortex slices) convincing evidence from kinetic data has now been obtained to justify the use of the inulin space as a measure of the true extracellular compartment, while sucrose and some other sugars as well as SO_4^{2-} may enter a compartment within the cells (Kleinzeller and Knotková, 1966).

3. Calculation of the Apparent Intracellular Ionic Concentration

Let $[A]_o$ be the concentration (mM) of the analyzed ion in the external medium; in the extracellular water $[A]_o$. H_2O_E mequiv. of this ion are then present. If A_t indicates mequiv. A/kg tissue DW, the apparent tissue concentration $[A]_t$ is A_t/H_2O_t.

The apparent intracellular ionic concentration $[A]_i$ (mM) is then $(A_t/H_2O_t - [A]_o \cdot H_2O_E)/H_2O_I$.

Comments

In the above calculation several assumptions were made:

a) The ionic concentration in the inulin space equals that in the medium;

b) all the intracellular water is available as solvent for the analyzed ion;

c) the ion is uniformely distributed in the intracellular water.

While the first assumption may be taken to be correct, evidence is available pointing against the justification of assumptions b) and c).

To test the availability of intracellular water as solvent, the equilibrium space for an easily permeable (i.e. of small molecular weight), non-dissociable, water-soluble, non-metabolizable and not actively transported substance may be determined. In our laboratory, ^{14}C-propanediol has been found to have the required properties for some animal tissues: in kidney cortex slices the space was found to be 97% (in rat diaphragm 74%).

The values for H_2O_I may then have to be corrected according to the found values of available intracellular water.

Assumption c) appears to be incorrect in the light of the steady-state efflux kinetics. In addition, a portion of the intracellular ions may be so slowly exchangeable with the labelled species that this fraction cannot be readily studied by efflux kinetics: This amount of intracellular ions may thus not participate in the osmotic and electrochemical processes between the cell and its medium. Thus, in rabbit kidney cortex slices about 86% of tissue Cl is rapidly exchangeable; however, in view of the distribution of Cl between the cell and the medium, the very slowly exchangeable fraction of Cl represents $\frac{1}{3}$ of the intracellular Cl (KLEINZELLER, NEDVÍDKOVÁ and KNOTKOVÁ, 1967).

II. Determination of Intracellular Cationic Activities Using Selective Microelectrodes

Problem. The measurement of activity of K^+ and H^+ ions in the cytoplasm of living skeletal muscle fibres.

Principle of Method. A microelectrode made of cation sensitive glass, previously calibrated in solutions with known cation activities, is introduced into the interior of frog sartorius muscle fibres. The potential difference between the cation electrode in the cytoplasm and a reference electrode in the medium is measured after subtraction of the membrane potential as measured with ordinary KCl filled microelectrodes. Calibration curves can be applied to convert the potential measurements with cation electrodes into ion activities.

Special Application. Measurement of potassium and hydrogen ion activities in frog muscle fibres in normal Ringer's solution.

Instruments. Microelectrode puller, special device for preparation of cation sensitive microelectrodes, high input resistance electrometer, micromanipulator, microscope. Plexiglass chamber for mounting muscle tissue, calomel reference electrode.

Solutions. 1.0, 0.1, 0.01 and 0.001 M KCl and NaCl (pH adjusted to 7.4 by tris buffer) phosphate buffer 0.01 M pH 5.0, 6.0, 7.0, and 8.0, methanol, 3 M KCl. Ringer's solution.

Theoretical Basis of Method. It is well known that thin glass membranes can be used as reversible electrodes for H ion determination. Only after systematic investigation of electrode behaviour of glasses of differing composition during the last 10 years [Schulz, 1953; Schulz and Ovchinikova, 1954; Schulz and Ayo, 1955; Eisenmann et al., 1957; Nicolsky et al., 1958 a, b, 1959; Isard, 1959; Materova et al., 1959, 1961; Savage and Isard, 1962; Eisenmann, 1962] did it become clear that some types of glass can be used for the determination of the activity of other cations besides H^+.

The electrochemical system used for the determination of cation activity is as follows:

$$
\begin{array}{c|c|c|c|c|c|c}
 & \text{I} & & & \text{II} & & \\
\text{Ag} & \text{AgCl; KCl; 3 M} & \text{Solution} & \text{Cation} & \text{KCl; 3 M} & \text{AgCl} & \text{Ag} \\
 & \text{satur.} & \text{examined} & \text{sensitive} & i^+; K^+, Cl^- & \text{satur.} & \\
 & & i^+, A^{n-} & \text{glass} & & & \\
 & & & \text{membrane} & & &
\end{array}
$$

$$\varphi_1 \qquad\qquad \varphi_{diff} \qquad\qquad \varphi_2 \qquad\qquad\qquad \varphi_3$$

The glass membrane is impermeable to anions and hence the Nernst equation may be used for determining the potential across it (φ_2)

$$\varphi_2 = \varphi_0' + \frac{RT}{F} \ln \frac{a_i^I}{a_i^{II}} = \varphi_0' + \frac{RT}{F} \ln a_i^I - \frac{RT}{F} \ln a_i^{II} . \tag{1}$$

The potentials φ_1 and φ_3 are equal but of opposite sign. Variations of the diffusion potential with variation of solution I can be neglected if concentrated KCl (where the mobilities of an- and cations are equal) is used at the liquid junction. The absolute value of this potential may be included in the standard potential φ_0 which, provided solution II does not change its composition, may also comprise $\dfrac{RT}{F} \ln a_i^{II}$. With these provisions a simple equation is obtained

$$\varphi = \varphi_0 + \frac{RT}{F} \ln a_i^I \tag{2}$$

where the potential is a linear function of $\ln a_i^I$.

The sensitivity of glass membranes is not strictly confined to one ion species. Depending on the composition of the glass, all ion species present in solution more or less affect the potential. Applying ion exchange theory to glass electrodes a general equation can be derived which describes the relationship between the potential and the activity of the various ions in the medium (NICOLSKY, 1937)

$$\varphi = \varphi_0 + \frac{RT}{F} \ln (a_i + k_{ij} \, a_j + k_{ik} \, a_k + \ldots + k_{iz} \, a_z) . \tag{3}$$

In this equation $a_i, a_j, a_k, \ldots a_z$ represent activities of the ions present in the solution and $k_{ij}, k_{ik}, \ldots k_{iz}$ represent the selectivity constants of the given glass.

The selectivity constants may be estimated by measuring the potential difference first in a solution containing species i only and subsequently in a solution containing some other cation, j, at equal activity $(a_i = a_j)$. It follows from Eq. (3):

$$\varphi_j - \varphi_i = \frac{RT}{F} \ln k_{ij}$$

$$k_{ij} = e^{\dfrac{F(\varphi_j - \varphi_i)}{RT}} \tag{4}$$

where φ_j and φ_i represent the electrode potentials measured in solutions containing only the jth or i^{th} cation species respectively.

Two methods can be applied to measure single ion activities in mixtures of unknown composition.

1. The use of highly selective glass electrodes for each cation of interest (to make the products $k_{ij}a_j, k_{ik}a_k, \ldots, k_{iz}a_z$ negligible compared to a_i);

2. the use of several glass electrodes with different known selectivities to obtain a system of equations

$$\varphi' = \varphi_0 + \frac{RT}{F} \ln (a_i + k'_{ij}a_j + \ldots + k'_{iz}a_z) \tag{5}$$

$$\varphi'' = \varphi_0 + \frac{RT}{F} \ln \left(a_i + k''_{ij} a_j + \ldots + k''_{iz} a_z \right)$$

$$\cdots\cdots\cdots\cdots\cdots\cdots\cdots\cdots\cdots \tag{5}$$

$$\cdots\cdots\cdots\cdots\cdots\cdots\cdots\cdots\cdots$$

$$\varphi = \varphi_0 + \frac{RT}{F} \ln \left(a_i + k_{ij} a_j + \ldots + k_{iz} a_z \right).$$

The first method may be used for the determination of the H^+ ion activity because many types of glass are highly selective for this ion (k_{HK} and $k_{HNa} > 10^{-7}$), and in some special cases for the determination

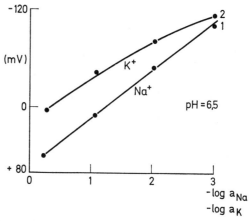

Fig. 1. Dependence of potential on $-\log a_{Na}$ (curve 1) and $-\log a_K$ (curve 2). Microelectrode made of NABS 25-05-09 glass

of other cations when the activity of one of them is much higher than that of all others.

In living cells both K^+ and Na^+ ions are present and for the determination of their activities it is better to use the second method. For the determination of intracellular potassium and sodium activities it is advisible to have microelectrodes with selectivity constants as different as possible; for instance, one electrode predominantly selective for potassium and the other for sodium ions. Unfortunately, at high K activities the behaviour of Na selective glass electrodes is not satisfactorily described by Eq. (3). As shown in Fig. 1 no linear relationship exists between potential and log a_K (curve 1) if a sodium selective glass electrode (made from NABS 25-05-09 glass) is used. It is therefore impossible to calculate the selectivity constant (k_{NaK}) for this case.

However, even such curves (whose shape depends on H^+ activities) can be fitted if, instead of Eq. (3), we use an empirical equation proposed by EISENMANN (EISENMANN et al., 1957).

$$\varphi = \varphi_0 + \frac{RT}{2F} \ln (a_H^{1/n_{HK}} + k_{HK}^{1/n_{HK}} a_K^{1/n_{HK}})^{n_{HK}} . \tag{6}$$

Alternatively, the equation for the 'generalized' ion-exchange theory of NICOLSKY and SCHULZ (1963) may be applied.

$$\varphi = \varphi_0 + \frac{RT}{2F} \ln (a_H + \bar{k}_{HL} a_K) + \frac{RT}{2F} \ln (a_H + \bar{k}_{HK} \alpha_{HK} a_K) . \tag{7}$$

The numerical values of the additional constants, n_{HK} of Eq. (6) and $\bar{k}_{HK}$ and α_{HK} of Eq. (7) may be derived from the dependence of the electrode potential on pH at constant a_K.

For curve 1 of Fig. 1 n_{HNa} Eq. (6) and α_{HNa} Eq. (7) are unity. For this special case, Eqs. (6) and (7) are similar to Eq. (3).

For precise activity measurements in cytoplasm with sodium selective electrodes one ought to employ Eq. (8) (BELJUSTIN and LEV, 1965; NICOLSKY et al., 1966).

$$\varphi = \varphi_0 + \frac{RT}{2F} (a_H + \bar{k}_{HK} a_K + \bar{k}_{HNa} a_{Na})$$
$$+ \frac{RT}{2F} (a_H + \bar{k}_{HK} \alpha_{HK} a_K + \bar{k}_{HNa} \alpha_{HNa} a_{Na}) . \tag{8}$$

It was already mentioned that α_{HNa} is very nearly equal to unity and thus, $\bar{k}_{HNa}$ of Eq. (8) is virtually the same as k_{HNa} of Eq. (3).

The system of equations to be solved for the precise determination of K^+ and Na^+ activities in cytoplasm is

$$\varphi_H = \varphi_0 + \frac{RT}{F} \ln a_H$$

$$\varphi_K = \varphi_0 + \frac{RT}{F} \ln (a_H + k'_{HK} a_K + k'_{HNa} a_{Na}) \tag{9}$$

$$\varphi_{Na} = \varphi_0 + \frac{RT}{2F} \ln (a_H + \bar{k}_{HK} a_K + k''_{HNa} a_{Na}) + \frac{RT}{2F} \ln (a_H +$$
$$+ \bar{k}_{HK} \alpha_{HK} a_K + k''_{HNa} a_{Na}) .$$

For the last method, determinations of the potentials of three different previously calibrated glass electrodes (φ_H, φ_K, φ_{Na}) are necessary. Measurements with pH-glass electrodes may be omitted only if the pH of cytoplasm could be estimated by some other method.

1. Construction of Cation Sensitive Glass Microelectrodes

A large number of different types of cation sensitive glass (for the composition and properties of electrode glasses, see EISENMANN et al.,

1957; NICOLSKY et al., 1958 a, b, 1961, 1962; EISENMANN, 1962, 1965, 1967) may be used for the construction of microelectrodes. Excepted are glasses with a low melting point, so called 'short glasses', and some very unstable glass mixtures. For the preparation of cation sensitive glass micro-electrodes, ordinary microelectrode pullers are used. If the glass is too soft (low melting point) it is difficult to find regimes for pulling electrodes with the required tip diameter $(0.5 - 0.7 \, \mu)$. Glasses with sharp transition from molten to solid state ('short' glasses) also cannot be used for the preparation of fine microelectrodes. Microelectrodes made of glass containing higher proportions of alkaline oxides (i.e. NAS 27-05) rapidly deteriorate when stored in aqueous solution or in moist air and therefore microelectrodes made from such glass must be used shortly after preparation. Nevertheless, glass like NAS 27-05 is one of the best for potassium selective microelectrodes. Selectivity of this glass is not very high $(k_{\mathrm{KNa}} = 0.08 - 0.12)$ but it behaves perfectly according to Eq. (3). For sodium selective microelectrodes NAS 11-18 and NABS 25-05-09 may be recommended. The composition of glasses is usually described as molar percent of oxides. Thus, glass NAS 11-18 has the composition Na_2O — 11 mol%, Al_2O_3 — 18 mol%, SiO_2 — 71 mol%; composition of glass NABS 25-05-09 is Na_2O — 25 mol%, Al_2O_3 — 5 mol%, B_2O_3 — 9 mol%, SiO_2 — 61 mol%. The potential of such electrodes measured in KCl containing solutions requires the application of Eqs. (6), (7), and (8).

The selectivity for NAS 11-18 is higher than for the NABS 25-05-09 glass but the nonlinearity of the potential dependence on ln a_{K} is also more pronounced.

For microelectrodes to be used in measurements of intracellular cation activities as designed by HINKE (1959) and further developed by LEV and BUZHINSKY (1961), the following conditions must obtain:

a) The tip of the electrode has to be fine so as not to damage significantly the cell membrane.

b) All the electrode except the tip has to be insulated to prevent the influence of cations in the extracellular fluid on the electrode potential.

The construction of microelectrodes begins with pulling very fine micropipettes on an ordinary microelectrode puller. Capillaries of cation-sensitive glass with an outer diameter of $1.0 - 1.2$ mm and inner one of $0.7 - 0.9$ mm serve as stock tubes for pulling inner micropipettes. The only special feature of the latter is the comparatively long 'neck' and small angle of its cone (Fig. 2).

Sealing of the tip of the cation sensitive inner micropipettes must be done immediately after pulling because of possible deterioration of the fine tip of certain cation sensitive (especially potassium selective) glasses. For sealing, the micropipette is fixed in a holder on the micro-scope table. Under microscopic control (objective 20 ×, eyepiece 15 ×)

a thin platinum wire loop is adjusted near the tip of the electrode. Then, by a controlled current, the loop is heated and as it elongates it comes into the close vicinity of the glass. It requires some skill and practice to seal the electrode channel without spoiling the fine tip.

After sealing, the electrodes are filled with 1 M or 3 M KCl in the usual way: Electrodes are first filled with methanol by boiling under reduced pressure; methanol is then replaced by distilled water and after 3 — 4 h the latter is replaced by KCl solution and the electrodes are kept there for 24 h to complete the diffusion process.

The insulating capillaries are prepared from Pyrex stock tubes (outer diameter of 2.0 mm and inner diameter of 1.5 mm) with an ordinary electrode puller. The puller is adjusted in such a way that the cone of the neck of the Pyrex micropipette becomes only a little larger than the neck cone of the inner pipette. The tip diameter of the insulating micropipette should be about 2 — 4 μ. The next step is to fill the tip of the insulation capillary with some insulating material of low melting point, e.g. a mixture of white picein wax with purified beeswax; for this, the tip is brought under microscopical control into contact with a drop of the melted wax mixture. The final step is the insertion of the inner micropipette into the outer one. This is done first by hand but when the tip of the inner pipette

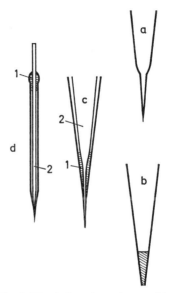

Fig. 2. The design of a cation sensitive microelectrode. a inner sealed micropipette made of cation sensitive glass; b Pyrex insulating pipette with a wax mixture in the tip; c cation sensitive glass covered with the insulation; d general view of cation sensitive microelectrode

reaches the neck of the outer one, the electrode is mounted on the holder of the microscope table. Using a micromanipulator, the inner pipette is then carefully pushed further just up to the wax plug. A heated platinum loop is then placed under the electrode tip and after the wax mixture has melted the inner micropipette tip is moved delicately through the wax to the outside of the insulation capillary. Heating is then stopped and a small drop of the same wax mixture is applied to the opposite end of the insulating sheath. The thin layer of wax covering the working tip of the cation sensitive micropipette may be removed either by short but comparatively strong heating with the platinum loop from a short distance which leads to evaporation, or by dipping the tip into ether for a short time.

The ohmic resistance of cation sensitive microelectrodes may be as high as $10^9 - 10^{12}$ and only some stable electrometers with an input resistance higher than 10^{13} (i.e. 'Vibron 33 B', Hungarian Electrometer — Type: TR-1501) may be used for measurements.

2. Calibration of Microelectrodes

Before measurement of intracellular activities, the microelectrodes must be carefully calibrated. The electrode pair (cation sensitive glass electrode and the reference electrode) are placed one by one in KCl and NaCl solutions of 1.0, 0.1, 0.01, and 0.001 M (pH $= 6.5 - 7.0$) and the potential difference for each solution is recorded. Using ordinary tables of mean activity coefficients and assuming them to be equal to activity coefficients of single ions ($\gamma_\pm = \gamma_+ = \gamma_-$), cation concentration may be converted into cation activities. Potential differences may then be plotted against $- \log a_K$ and $- \log a_{Na}$ and in general a straight line should be obtained with a slope (S) near the theoretical value of $\dfrac{RT}{F} =$ 57 mV at room temperature; by extrapolation to the value of $\log a = 0$ ($a = 1$) the standard potential φ_0 may be found. If the slopes (S) of the calibration curves for KCl and NaCl solutions are the same, one may calculate the selectivity constants from the potential difference at any activity value and magnitude of the slope:

$$k_{KNa} = e^{\dfrac{-\Delta\varphi}{S}}. \tag{10}$$

It was shown above that for sodium glass microelectrodes, the calibration curve as measured in KCl solution is not linear; for a precise determination of intracellular activities Eq. (8) should be used. To apply this equation, the constants $\bar{k}_{HK}$, α_{HK}, and k_{HNa} must be known. Here an additional calibration curve with three or more tris chloride solutions of different pH are used. The constant k_{HNa} is easily calculated in the usual way from the difference between potentials for any $a_{Na} = a_H$. If there is no overlap of the range of the calibration curves for Na^+ and H^+ the two curves may be linearly extrapolated since for these ion species they are practically always linear. The constants $\bar{k}_{HK}$ and α_{HK} may also be calculated from these data. Calibration of pH microelectrodes is simple. Only several precisely adjusted buffer solutions are required (Caldwell, 1954, 1948; Sorokina, 1961; Kostyuk and Sorokina, 1961).

3. Intracellular Measurements

After calibration, the cation sensitive electrode is ready to be inserted into the cell. It is possible to impale the cell simultaneously with a cation sensitive electrode as well as with a conventional open microelectrode. This technique is applicable only if the tips of both electrodes can be positioned near to each other. It is simpler to measure first the membrane potential with a conventional microelectrode and then the potential difference between the cation sensitive microelectrode inside the cell and the reference electrode connected with the medium [LEV, 1964 (1, 2)]. This latter potential difference includes the membrane potential of the cell which must be substracted.

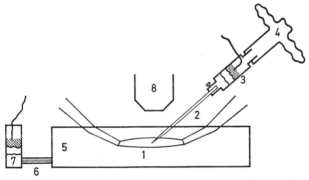

Fig. 3. The arrangement for the measurement of intracellular ionic activities. 1 muscle fixed in forceps; 2 cation sensitive microelectrode connected with calomel electrode (3) in the holder of the micromanipulator (4); 5 bath with Ringer solution connected by capillary (6) to calomel reference electrode (7); 8 microscope objective

Procedure for Measurement in Frog Muscle Fibres. The muscle is mounted in a bath containing frog's Ringer solution. A silver-silver chloride or calomel reference electrode is connected through a very fine capillary with the medium. Cation sensitive and ordinary microelectrodes, also connected to the same electrodes are mounted in the holder of a micromanipulator. The Fonbrune type was found to be useful for impaling cells under microscopic control at magnification of $200 - 330 \times$ (Fig. 3).

First, the resting potential is measured in $10 - 20$ cells. Cation sensitive electrodes are then introduced in regular sequence into $10 - 20$ cells. The difference between the mean values of the potential of the cation sensitive microelectrodes and the mean membrane potential is used to determine the ionic activities from the calibration curves [LEV and BUSHINSKY, 1961; LEV, 1964 (1, 2); SOROKINA, 1961].

If there is reason to believe that the ratio of intracellular K^+ and Na^+ activities $(a_K : a_{Na})$ is not high enough for a precise determination of a_K

sodium selective microelectrodes (or also potassium selective electrodes, but their selectivity constant must be significantly different from that of the first electrode) should be used.

For determination of sodium activity the two-electrode method is the method to choice [LEV, 1964 (1, 2)].

The determination of intracellular pH may be carried out with a single pH microelectrode provided this electrode has been previously checked for the absence of salt error (CALDWELL, 1954, 1958; SOROKINA, 1961; KOSTYUK and SOROKINA, 1961); or the intracellular pH may be determined by some other method (see comments below).

Comments

A. Possible error of the glass microelectrode method

a) One probable error originates from an uncertainty of the change of the diffusion potential of the open microelectrode (reference electrode) when the latter is transferred from the medium into the cytoplasm. This error is present in all determinations of membrane potentials. It may be partly avoided by careful selection of such open electrodes which do not significantly change their potential when the cation activities in the bathing solution are changed. Theoretically, an error of this kind may have some influence on the absolute values of cation activities but the sequences of activities $(a_i : a_j : a_k : \ldots : a_z)$ found by this method are not subject to that error [LEV, 1964 (1, 2)].

b) Errors of another kind appear if the value of φ_0 or the selectivity constant of the electrode change at the time of measurement. To avoid this error a second calibration of the electrode should be carried out after intracellular measurements. The stability of the electrode potential as measured between two impalements in the medium may also be used as evidence for unchanged electrode properties.

c) Even if prepared with great skill, the size of the tip of cation sensitive microelectrodes varies and thus some differences in the magnitude of membrane potential may be expected. Naturally, the magnitude of this difference is proportional to the value of the membrane potential itself. This error may be minimized by depolarizing the membrane (e.g. by substituting a part of the NaCl in the medium by K_2SO_4) without significant change of the intracellular cation activities.

B. pH determination

As an example of a chemical method for intracellular pH determination the following procedure may be recommended (WADDEL and BUTLER, 1959; KOTYK, 1963):

A weak acid dye such as bromphenol blue (pK = 4.0) distributes itself between medium and cell interior until its undissociated form reaches equilibrium. The intracellular pH can be computed from the following formula:

$$pH_i = pH_o + \log\left[\frac{c_i}{c_o}(1 + 10^{pK-pH_o}) - 10^{pK-pH_o}\right]. \tag{11}$$

The intracellular concentration of the dye c_i and the extracellular one c_o are determined from the decrease of the concentration in the medium as measured by simple colorimetry. The amount bound to cell constituents (estimated in a sonicated cell suspension) must be substracted and the solution has to be made alkaline before colorimetry in order to obtain complete conversion of the dye to the coloured form.

N.B. The use of 5.6-dimethyl-2.4-oxazolidindione which is frequently recommended for animal cells is not suitable since the dye may be actively transported.

C. Interpretation of results obtained by the methods described in this section

The measured values of intracellular activities for Na^+ and K^+ are subject to some uncertainties. The sensitive tip of the microelectrode may be in contact with one or several intracellular compartments, e.g. with the hyaloplasm itself, with the contents of the endoplasmatic reticulum, etc.; thus, the measured values will represent either the activity in one impaled compartment or a compound activity for several compartments.

Analoguous considerations apply to the estimation of intracellular pH with microelectrodes and thus differences may arise between values obtained by microelectrodes and chemical methods.

References

BACON, J. S. D., and D. BELL: Biochem. J. **42**, 397 (1948).

BELJUSTIN, A. A., and A. A. LEV: In: Chemistry in natural sciences. Leningrad: University, USSR 1965.

CALDWELL, P. C.: J. Physiol. (Lond.) **126**, 169 (1954).

— J. Physiol. (Lond.) **142**, 22 (1958).

EISENMANN, G.: Biophys. J. **2**, part 2, 259 (1962).

— Adv. anal. chem. Instr. **4**, 213 (1965).

—, D. O. RUDIN, and J. CASBY: Science **126**, 831 (1957).

—, Ed.: Glass electrodes for hydrogen and other cations. New York: Marcel Dekker 1967.

HINKE, J. A. M.: Nature (Lond.) **184**, 1257 (1959).

HÖFER, M.: Experientia (Basel) **19**, 367 (1963).

ISARD, J.: Nature (Lond.) **184**, 1616 (1959).

KLEINZELLER, A.: In:Manometrische Methoden und ihre Anwendung in der Biologie und Biochemie. Jena: G. Fischer-Verlag 1965.

—, and A. KNOTKOVA: Biochim. biophys. Acta (Amst.) **126**, 604 (1966).

—, NEDVIDKOVA, J., and A. KNOTKOVA: Biochem. biophys. Acta (Amst.) **135**, 286 (1967).

Kostyuk, P. G., and Z. A. Sorokina: In: Kleinzeller, A., and A. Kotyk, Eds., Membrane transport and metabolism, a Symposium. Prague: Publ. House of the Czechosl. Acad. Sci. 1961.

Kotyk, A.: Folia microbiol. (Praha) 8, 27 (1963).

Kulka, R. G.: Biochem. J. 63, 542 (1956).

Lev, A. A.: (1) Nature (Lond.) 201, 1132 (1964).

– (2) Biophysica (USSR) 9, 686 (1964).

—, and E. P. Buzhinsky: Cytology (USSR) 3, 614 (1961).

Materova, E. A., V. V. Moiseev, and A. A. Beljustin: J. Phys. Chem. (USSR) 35, 1285 (1961).

— —, and S. P. Schmidt-Fogelevich: J. Phys. Chem. (USSR) 33, 893 (1959).

Nicolsky, B. P.: J. Phys. Chem. (USSR) 10, 495 (1937).

—, M. A. Schulz, and A. A. Beljustin: Dokl. Akad. Nauk. (USSR) 144, 844 (1962).

— —, and N. V. Peshehonova: (1) J. Phys. Chem. (USSR) 32, 19 (1958).

— — — (2) J. Phys. Chem. (USSR) 32, 262 (1958).

— — — J. Phys. Chem. (USSR) 33, 1922 (1959).

— —, A. A. Beljustin, and A. A. Lev: In: G. Eisenmann, Ed., Glass electrodes for hydrogen and other cations. New York, Marcel Dekker 1967.

— —, E. A. Materova, and A. A. Beljustin: Dokl. Akad. Nauk. (USSR) 140, 641 (1961).

Ramsey, J. A., R. H. J. Brown, and P. C. Crohgan: J. exp. Biol. 32, 822 (1955).

Savage, J. A., and J. Isard: Phys. and chem. Glasses 3, 142 (1962).

Schulz, M. M.: Uchenye zapiski Leningrad University, No. 169. Ser. Chem. Issue 13, 80 (1953).

—, and L. G. Ayo: Vestn. Leningrad Univ. No. 8, p. 153 (1955).

—, and T. M. Ovchinikova: Vestn. Leningrad Univ., No. 2, p. 129 (1954).

Sorokina, Z. A.: Cytology (USSR) 3, 48 (1961).

Umbreit, W. W., R. H. Burris, and J. F. Stauffer: Manometric techniques. Minneapolis: Burgess Publ. Co. 1957.

Wadell, W. J., and T. C. Butler: J. clin. Invest. 38, 720 (1959).

Experiments on Na Transport
of Frog Skin Epithelium

By T. W. CLARKSON and B. LINDEMANN

I. Introduction

a) Purpose. The experiments were chosen for their educational value and have two objectives. First, active transport will be demonstrated in terms of the short circuit current (SCC). The ionic species transported will be identified by comparing the SCC with the net flux of sodium as measured by tracer techniques. Second, hormonal influence of active transport (SCC) will be observed. The hormonal effect will be analyzed in more detail in terms of a compartmental model of frog-skin.

According to this model, the frog skin is regarded as a single compartment or cell layer separating an "inner[*]" and an "outer" bathing solution. The transport of sodium ions across frog skin involves the movement of sodium across two membranes — the membrane facing the outer solution and that facing the inner solution. Sodium diffuses passively across the outer membrane but is actively transported out of the cell across the inner membrane. As a result one observes a net transport (flow) of sodium across the isolated frog skin from the outer to the inner bathing solution.

This net flow of sodium across each membrane of the skin is composed of two unidirectional fluxes. Thus for the outer facing membrane, the inward flux is M_{12} and the outward flux is M_{21}. The net flow of sodium J_o across the membrane is the difference between M_{12} and M_{21}

$$J_o = M_{12} - M_{21} . \tag{1}$$

Similarly the net flow of sodium J_i across the inner membrane is the difference of the two fluxes M_{23} and M_{32}

$$J_i = M_{23} - M_{32} \tag{2}$$

Compartment 1	Compartment 2	Compartment 3
Outer solution	Frog skin	Inner solution
$\longrightarrow M_{12}$	$M_{32} \longleftarrow$	
$\longleftarrow M_{21}$	$M_{23} \longrightarrow$	

[*] The term "inner" refers to solutions in contact with the anatomically inner surface of the skin.

When the net flow of sodium from outer to inner solution (J) has become steady, the net flow of sodium into the cell is equal to the net flow leaving the cell i.e.

$$J = J_o = J_i \qquad (3)$$

and

$$M_{12} - M_{21} = M_{23} - M_{32} .$$

Three fluxes (M_{12}, M_{21}, M_{32}) represent passive movement of sodium whereas M_{23} arises from both active and passive processes.

Measurement of the tracer movement in the approach to a steady tracer flux and in the steady state allow calculation of the four unidirectional fluxes and the net flow of sodium.

b) The Technical Problem

1. To ensure that the tissue is in the steady state (net sodium flux is constant).

2. To measure the unidirectional fluxes with ^{24}Na.

1. Steady-state. In the isolated frog skin, the sodium transport may be measured in two ways — by use of the ^{24}Na isotope and electrically. We shall use both methods so to check one against the other and also use the electrical method to check that the tissue has achieved a steady state. The electrical measurement of net sodium flow across the skin is based on the following principle. Usually the isolated frog skin develops an appreciable voltage difference — about 50 mV or more. This may be reduced to zero by applying an external current — usually called the short-circuit current. In the short-circuited condition, when the isolated frog skin is bathed by identical solutions, no external driving forces are acting to cause passive movement of ions. Thus the short-circuit current is equal to the algebraic sum of all the flows of ions which are transported actively. In the case of the frog skin under the given experimental conditions, sodium is the only ion undergoing active transport. The short-circuit current, therefore, gives a continuous record of net sodium flow across the frog skin and offers a convenient means of checking that the tissue is in the steady-state.

2. Fluxes of ^{24}Na. In the special lucite chambers provided, the skin area is divided into two isolated halves designated 1 and 2, each having its own separate half-chambers which contain the "inside" and "outside" bathing solution. An external current can be passed across each half of the skin to allow measurement of the short-circuit current. The isotope fluxes are measured by adding ^{24}Na to a solution bathing one side of the skin and measuring its appearance on the other side of the skin as a function of time. A sample of ^{24}Na is added to the appropriate bathing solution of the other skin half to measure the flux in the opposite direction.

The hormone is added to the two *inner* subchambers, since it cannot reach its site of action from the outside solution.

During the first 15 min after the addition of ^{24}Na, the fluxes across the skin are not steady because the isotope is mixing with unlabelled sodium inside the skin. Thus, although the tissue is in steady state with regard to the total sodium, the isotope is approaching steady state.

As shall be described in the mathematical section later, the four unidirection fluxes M_{12}, M_{21}, M_{23} and M_{32} are computed from the kinetics of tracer flow (a) in the approach to steady state (first 15 min) and (b) in the steady-state (say 1 h after addition of isotope). It will also be necessary to measure the steady-state ^{24}Na sodium content of the skin, by counting the radioactivity of the skin at the end of the experiment.

II. Instruments and Solutions

Frogs, dissection equipment, Ringers solution and glassware, four compartment lucite chamber, agar bridges, calomel and Ag-AgCl electrodes, air supply and Krogh pumps, gas washing flasks and fluid traps, plastic tubing, mV-meter, short-circuit apparatus and μA-meter, clock, thermometer, Na^{24} in Ringers solution, automatic pipettes and glassware for adding isotope and sampling, well type counting equipment, decay chart, lead bricks, rubber gloves, plastic buckets.

Lysine-8-Vasopressin is used in a final concentration of 0.2 I.U./ml.

Composition of Ringers solution: NaCl: 115 mM/l, NaHCO$_3$ 2.4 mM KCL 2.0 mM. Reprints of original publications related to these experiments will be supplied.

III. Description of Experiment

Large frogs of the species Rana esculenta will be used. The frogs are killed, the abdominal skins isolated and mounted on the apparatus. This procedure will be demonstrated. All the air leads to the Krogh pumps are sealed tight and exactly 10 ml of Ringers solution added to each sub-chamber. Start the air-flow through the Krogh pumps. Open the air leads cautiously! Adjust the rate of bubbling to give as fast a flow as possible without splashing. Record the open-circuit voltage across each skin for about 5 min. Apply current to reduce the voltage to zero. This is the short-circuit current (SCC). Read this at intervals of approx. 5 min and record the value. Do this for the duration of the experiment. A typical recording is given in Fig. 3.

While waiting for the short-circuit current* to become steady, (about 60 min from short-circuiting), rehearse the sampling technique.

---

* Approximately 30 min after short-circuiting, and on judgement of the instructor, ADH will be added to both inside solutions of every second experimental set up.

A schematic of the short-circuit apparatus is shown in Fig. 1.

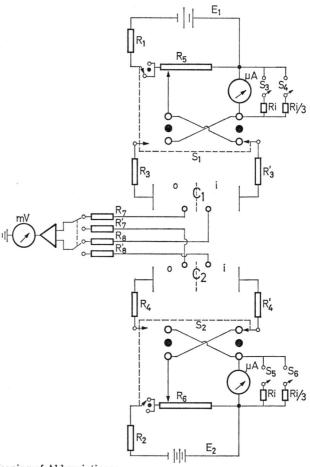

Fig. 1. Meaning of Abbreviations:

$E_{1,2}$	90 volt anode batteries
$R_{1,2}$	current limiting resistors 12 K ohm
$R_{3,4}$ $R'_{3,4}$	resistance of AgAgCl cells plus agar KCl bridges
R_i	resistance equal to internal resistance of current meter
$R_{i/3}$	resistance equal to one third internal resistance of current meter
$R_{5,6}$	Helipots to adjust current through skin
R_7, R'_7, R_8, R'_8	Potential electrodes (calomel) plus agar KCl bridges
$S_{1,2}$	three position switches for
	1. open circuit
	2. inward current
	3. outward current
$S_{4,5,6,7}$	switches to adjust scale of current meter by 2 × and 4 ×
$C_{1,2}$	two subchambers for the two parts of skin area — each subchamber consists of an outer (o) and an inner (i), separated by 4.7 cm² skin area (dotted line). The vertical solid lines represent the rear of the chambers where the current is introduced.

The chamber arrangement is schematically shown in Fig. 2.

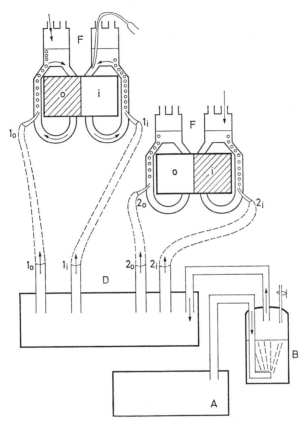

Fig. 2. Meaning of Abbreviations:

A   compressed air supply

B   device to moisten gas

D   a set of Hoffman clamps to regulate flow of gas, and fluid traps to make flow
        of solutions between half chambers or subchambers impossible

$Ch_{1,2}$ two subchambers, each consisting of an outer and inner halfchamber,
        4.7 cm² of exposed skin in each subchamber

1_o, 1_i)   gas inlets of Krogh pumps
2_o, 2_i)

F   special reservoirs used for adding insotope and taking samples. Total fluid
        volume in each half chamber plus reservoir is about 10 ml. The chamber
        arrangement is mounted behind a wall of lead bricks.

It will first be demonstrated by the instructor. All sampling will be done by one student only. The other will act as the "scribe", noting time, sample numbers, SCC etc. When the SCC is steady and you are satisfied that the sampling is accurate (not more than 1% error), the instructor will add a small volume of ^{24}Na to the inside chamber 2 (2_i) noting the clock time to the nearest min. Then ^{24}Na is added to the outside chamber 1 (1_o) and the stop watch started. Samples are now taken from chamber (1_i) at approximately 1.5 min intervals for the next 15 min. The time of collec-

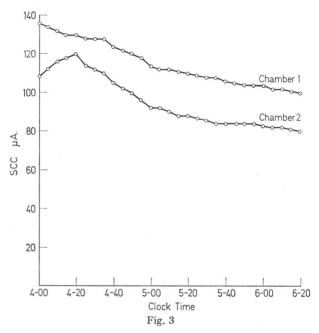

Fig. 3

tion is noted to the nearest second. The sample volume is 0.5 ml in all cases. Immediately after withdrawal of each sample, volume is replaced by addition (using the same syringe) of 0.5 ml of non-radioactive Ringers solution. The samples are placed in the counting tubes provided. The student taking the samples should avoid touching these tubes with his gloves. 60 min after addition of isotope, sampling is continued from (1_i) and started from (2_o) in 15 min intervals for the next 75 min. At the end of this period, 0.5 ml samples are taken from the two "hot" sub-chambers (1_o and 2_i) and diluted into 100 ml of saline in the volumetric flasks provided. $\frac{1}{2}$ ml of the diluted solution is added to the counting tubes. All media are carefully removed from the chambers and the volumes recorded. The instructor will explain how this is done. The two halves of the skin exposed to the bathing solutions are carefully cut out, dipped in non-radioactive Ringers solution for exactly 5 sec (use stop watch) and rolled in parafilm. The parafilm is folded into a small package and placed at the bottom of the counting tube. Please carefully identify each half of the skin (the one exposed to ^{24}Na on the outside is No. 1 and the other is No. 2).

The area of skin exposed in each sub-chamber is 4.7 cm^2.

For listing and computation of the raw data, the following charts may be used (Table 1).

IV. Theory, Computation and Display of Data

a) Comparison of the Net Flow of Sodium with the SCC

The sodium flux M_{13} from the outer to the inner solution is computed from the equation

$$M_{13} = \frac{S_1}{P_1} \cdot \dot{P}_{300} \text{ equiv./min}$$

where the symbols are defined as

S_1 = the total sodium content in equivalents of subchamber 10,
P_1 = the total ^{24}Na content in counts/min of subchamber 10,
$\dot{P}_{300}$ = the rate of increase in the amount of ^{24}Na in subchamber 1$_i$ in counts · min^{-2} when the isotope flux has become steady.

Similarly the sodium flux in the opposite direction is computed from the equation

$$M_{31} = \frac{S_3}{P_3} \cdot \dot{P}_{100} \text{ equiv./min} .$$

The difference between M_{13} and M_{31} is the net flow of sodium.

The short-circuit SCC is equal to the algebraic sum of the net flows of all ions moving across the skin. If sodium is the only ion transported under the conditions of these experiments, the net flow of sodium J_{Na} must be equal to the SCC.

$$\text{i.e. } J_{Na} = M_{13} - M_{31} \approx \text{SCC} .$$

The SCC, measured in µA, is converted to the same units as J_{Na} (equiv./min) as

$$1 \text{ µA} = 1 \text{ µ coul/sec} = 60 \text{ µ coul/min}$$

$$= \frac{60 \text{ µ}}{96,500} \text{ equiv./min} = \frac{60 \times 10^{-6}}{96,500} \text{ equiv./min}$$

where 96,500 is Faraday's Number.

Table 1

Medium	Sample	Time of sampling h min	SCC μA	Time of begin of counting h min	Correct. factor	Count. time for 20000 c	cpm for 0.5 ml − Bgr	cpm dec. corr.
Background (Bgr)	0							
1_i	1							
	2							
	...10							
1_i	11							
	...17							
	18							
$1_o \left(\frac{1}{100}\right)$	19							
Skin area 1	20							
2_o	...27							
$2_i \left(\frac{1}{100}\right)$	28							
Skin area 2	29							

The total amount of radioactivity which has entered medium (1_i) or (2_o) up to the time of sampling of the k'th sample is given by

$$P_k = [2 \cdot r_k \cdot V + \sum_1^{k-1} r_i]$$

where r_k is (cpm/0.5 ml) of the k'th sample and V is total volume of medium (10 cm³).

b) Discrimination between Passive and Active Transport

Following the treatment of Ussing [1] and Schlögl [2] we write for simple diffusion

$$\frac{M_{13}}{M_{31}} = \frac{\bar{a}_1}{\bar{a}_3}$$

where M's are unidirectional fluxes and $\bar{a}$'s are electro-chemical activities of the solute in the external media:

$$\bar{a}_{i1} = c_{i1} f_{i1} \exp [\psi_1 \, z_i \, F/RT] \, .$$

Taking logarithms of both sides and making use of

$$\mu_{i1} = \mu_{oi1} + RT \ln \bar{a}_{i1} \, , \qquad \mu_1 - \mu_3 = RT \ln (\bar{a}_1/\bar{a}_3)$$

we obtain Ussings Eq. (3):

$$RT \ln (M_{13}/M_{31}) = \mu_1 - \mu_0 \, .$$

This relationship can be used to prove the absence of other than simple diffusion mechanisms for a steady state transport system. For instance, by keeping $(\mu_1 - \mu_3)$ zero, (absence of chemical, electrical and hydrostatic gradients), we expect

$$M_{13}^{Na} = M_{31}^{Na}$$

in the steady state, if simple diffusion is the only mechanism responsible for the Na fluxes.

Conversely, the observation under these conditions of a steady SCC is proof of "active transport" of one or more ionic species, and the observation of a steady net-flux of Na is proof of active Na transport.

In general we can be sure that active transport is responsible for an observed flow of solute when we have excluded other driving forces, such as electrochemical gradients, or bulk flow (1). This definition implies that the molecular machinery of the cellular membrane can convert free energy of protoplasmic constituents into transport work.

It is suggested that the literature quoted be consulted for details.

c) Rate Constants and Pool Size

We shall follow in general the treatment of Curran et al. [4, 5], which is a modification of that of Schoffeniels [6] and of Hoshiko and Ussing [7]. These four papers should be consulted for details. Compare also Ref. [22].

The epithelium is considered to behave as a single compartment (index 2), which is bound by different barriers for Na diffusion towards the "outer" (index 1) and the "inner" (index 3) external media. We thus obtain a system of three compartments in series with four rate constants (Fig. 4). (For the time being we neglect the presence of the corium.)

When P_i denotes total radioactivity in compartment i:

$$\dot{P}_2 = - (k_{21} + k_{23})\, P_2 + k_{12}\, P_1 + k_{32}\, P_3\,.$$

Assumptions: a) All quantities connected with compartment 1 are constant,
  b) all isotope rate constants (k) are in fact constant,
  c) V_2 (Volume) is constant (Tissue in steady state),
  d) P_3 is close to zero and can be neglected,
  e) $P_{2\infty}$ can be measured by counting the whole skin and applying a correction for the isotope content of the extracellular space★.

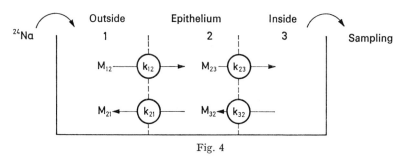

Fig. 4

Thus

$$\dot{P}_2 + k\, P_2 - k_{12}\, P_1 = 0$$

where

$$k = k_{21} + k_{23}$$

since $P_2 = 0$ when $t = 0$, we obtain

$$P_2 = \frac{k_{12}}{k} \cdot P_1(1 - e^{-kt})\,.$$

In the steady state with respect to tracer flow, $\dot{P}_2$ is zero, and

$$P_2 = P_{2\infty} = P_1\, k_{12}/k$$

therefore

$$P_2 = P_{2\infty}(1 - e^{-kt})$$

for all values of t.
  Equally

$$P_3 = k_{23}\, P_2 = k_{23}\, P_{2\infty}\, (1 - e^{-kt})$$

★ The corial connective tissue layer between compartments 2 and 3 does not seriously lower the rate of tracer appearance in compartment 3 except in the first few minutes after addition of tracer (4). However, the connective tissue layer gives rise to a larger extra-cellular space on this side of the tissue. Consequently only the tissue whose outer surface is exposed to the solution of high radioactivity may be used to compute $P_{2\infty}$.

which becomes

$$\dot{P}_{300} = k_{23} \, P_{200} = \text{constant}$$

for large t. Further transformations yield

$$\ln(1 - \dot{P}_3/\dot{P}_{300}) = - \, kt \, .$$

Thus $k = k_{23} + k_{21}$ can be obtained by plotting $\ln[1 - (\dot{P}_3/\dot{P}_{300})]$ versus time. The individual rate constants are evaluated as follows:

$$k_{12} = k \, P_{200}/P_1$$

$$k_{23} = \dot{P}_{300}/P_{200}$$

$$k_{21} = k - k_{23} \, .$$

P_{200} is obtained, at least approximately, by counting the exposed skin area itself at the end of the experiment and subtracting 35% of the total skin count. (Correction of CEREIJIDO et al. for extracellular space [5].)

In the steady state, the net flux of Na is

$$\text{SCC}/F = k_{12} \, S_1 - k_{21} \, S_2$$

where S is the total Na content. Therefore

$$S_2 = \frac{k_{12} \, S_1 - (\text{SCC}/F)}{k_{21}}$$

is the Na pool of the compartment. Having computed this quantity, we can try to find the remaining rate constant:

$$k_{32} = \frac{k_{23} \, S_2 - (\text{SCC}/F)}{S_3}$$

The four rate constants and S_2 should be computed from the data. You should compare your results with those obtained by other groups to evaluate the effects of ADH.

Since the rate constants are now known, we can proceed to compute the unidirectional Na fluxes through each barrier for the steady state, using the relationship

$$M_{ij} = k_{ij} \cdot S_i \, .$$

d) Details of Computations

The following steps should be carried out:

1. Convert all counts to counts per min (cpm).

2. Subtract background (cpm).

3. Correct for decay using the table provided. Correct all the counts to one common point in time, the most convenient being the time that the first sample was counted.

4. Counts on samples taken from the bathing solutions, should be multiplied by the appropriate dilution factors to give the total cpm in the

solutions (i.e. P_1 and P_3). The dilution factor for samples from the hot solutions (chambers 1_o and 2_i) is 4000 and 20 for the other samples (chambers 1_i and 2_o) assuming the volumes are exactly 10 ml.

5. Correct the values of P_1 and P_3 for the 0.5 ml of solution removed in sampling. — See equation in the previous text. Thus for any value P_k, the correction involves adding the sum of all the *previous* 0.5 ml sample counts.

6. To calculate $P_{2\infty}$ take the count of that half of the tissue where outer surface was exposed to the high ^{24}Na solution (half No. 1) and reduce this value by 35%.

7. Divide $P_{2\infty}$ by P_1 (the counts in chamber 1_o, which are always constant) to obtain the value of k_{12}/k.

8. P_3 is the total cpm in chamber 1_i. Calculate the increase in P_3 during the 15 min sampling periods in the last 75 min of the experiment. These values of ΔP_3 should be constant (show no definite trend to increase or decrease but they may fluctuate due to errors in sampling and counting). Take the mean value of ΔP_3 and divide by 15 to give $\dot{P}_{3\infty}$ in cpm/min.

9. Now calculate $\dot{P}_3$ from the ratio $\Delta P_3/\Delta t$ where ΔP_3 is the increase during the first 15 min after addition of isotope and Δt is the time between collections of successive samples in minutes.

Plot the function $\ln(1 - \dot{P}_3/\dot{P}_{3\infty})$ against time in minutes. The gradient of the line is $- k$ (see Fig. 5).

You may use the semi-log paper provided but note this paper is in units of log to the base 10 and $2.303 \log_{10}k = \ln k$, so the gradient must be corrected.

10. Since the value of k_{12}/k is known, k_{12} may be calculated.

11. Calculate k_{23} from the equation $k_{23} = \dot{P}_{3\infty}/P_{2\infty}$.

12. Calculate k_{21} from the equation $k = k_{21} + k_{23}$.

13. From your graph of SCC against time, calculate the mean value of the SCC over the time period from addition of isotope to the end of the experiment. Do this for both halves of the tissue. These values should be close in the ideal experiment. If not, use the mean SCC from skin half/ No. 1 in subsequent calculations.

14. Convert SCC in μA to μ equiv./min.

15. Express S_1, the total sodium in bathing solution of sub-chamber 1_o, in μ equiv.

16. Calculate S_2 the sodium content of the skin, from the equation

$$\mathrm{SCC} = k_{12} S_1 - k_{21} S_2$$

where SCC is in μ equiv./min, S_1 is in μ equiv. and k_{12} and k_{21} are in min^{-1}.

17. Calculate S_3 in μ equiv. In this experiment, $S_1 = S_3$.

18. Calculate the remaining rate constant k_{32} from the equation

$$k_{32} = \frac{k_{23} S_2 - \text{SCC}}{S_3} \; .$$

19. Since all four rate constants are now known, we can calculate the unidirectional sodium fluxes through each barrier in the steady-state from the equation

$$M_{ij} = k_{ij} S_i \; .$$

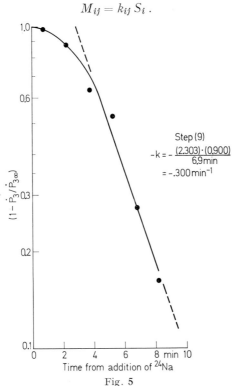

Step (9)

$$-k = -\frac{(2.303) \cdot (0.900)}{6.9 \, \text{min}}$$

$$= -.300 \, \text{min}^{-1}$$

Time from addition of ^{24}Na

Fig. 5

20. The area of each half of exposed skin is 4.7 cm². You may correct the observed values of M_{12}, M_{21}, M_{23} and M_{32} to fluxes per unit area.

Fig. 6 gives a "flow sheet" of the calculations from the measured values to the calculated constants. It indicates that certain constants e.g. k_{23} are calculated directly from only two observations whereas others e.g. k_{32} are estimated from several observations and from other calculated constants. It also indicates that certain observations are used more frequently than others in the calculations. For example, the tissue count $P_{2\infty}$ is involved in practically every calculation and so must be measured with great care.

7   Membrane Biophysics

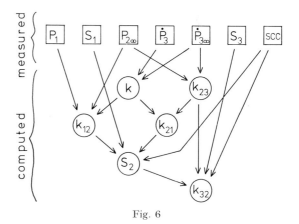

Fig. 6

e) Display of Compartmental Model of the Frog Skin by Analog Computer.

Fig. 7A shows a patch diagram of the general case of a catenary system of compartments. The diagram below (Fig. 7B) deals with the special case of 3 series compartments, which is used in this paper for analysis of frog skin data.

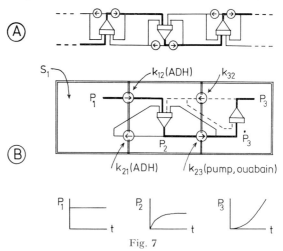

Fig. 7

A small analog computer can be patched according to Fig. 7B to display changes in P_1, P_2, and P_3 with time. In the training course of 1966a Telefunken RA 741 was used for this purpose[*]. The effect of ADH (increase k_{12} and k_{21}) and of ouabain (decrease k_{23}) can easily be demonstrated with the computer.

[*] Made available by the Volkswagenwerkstiftung, to which our thanks are due.

To ease comparison with Fig. 4, the analog computer patch diagram of Fig. 7B has been drawn into the framework of a system of three series compartments, such that the potentiometers (circles) which represent adjustable rate constants of transport through membranes appear projected onto these membranes (double lines). Non-membrane compartmental boundaries are also represented by doubles lines. Note that the size of the inner (epithelial) compartment (2) is in reality much smaller than that of the boundary compartments (1 and 3). Dashed connections carry signals of negligible amplitude.

V. Brief Survey of Literature

The short circuit current-method is the easiest and most direct way of demonstrating the phenomenon of "active transport". It has been successfully applied to a variety of epithelia and cell membranes since its introduction by Ussing and Zerahn in 1951 [3]: to epithelia of stomach [8], small [9, 10] and large intestine [11, 12] urinary bladder [13], kidney tubules [14], cornea [15] and ciliary body [16]. While it is often easy to demonstrate a SCC it may be difficult to identify the ionic species transported.

In many cases more than one ionic species seems to be involved, and in some epithelia one can not yet account for large parts of the SCC. It should be noted that the short circuit technique is a special case of the voltage clamp technique (with zero command potential) which, since its introduction to membrane research [17, 18, 19] has been equally successful.

For introduction to present day problems of research on epithelial permeability consult for instance [10 − 35], where further reference may be found.

VI. References

1. Ussing, H. H.: Acta physiol. scand. 19, 43 (1949).
2. Schlögl, R.: Habilitationsschrift. Göttingen 1957.
3. Ussing, H. H., and K. Zerahn: Acta physiol. scand. 23, 110 (1951).
4. Curran, P. F., F. C. Herrera, and W. J. Flamigan: J. gen. Physiol. 46, 1011 (1963).
5. Cereijido, M., F. C. Herrera, W. J. Flamigan, and P. F. Curran: J. gen. Physiol. 47, 879 (1964).
6. Schoffeniels, E.: Biochem. biophys. Acta (Amst.) 26, 585 (1957).
7. Hoshiko, T., and H. H. Ussing: Acta physiol. scand. 49, 74 (1960).
8. Rehm, W. S., and W. H. Dennis: Metabolic aspects of transport across cell membranes. U. of Wisconsin Press 1957.
9. Schultz, S. G.: J. gen. Physiol. 47, 567, 1043 (1964); 48, 375 (1965).
10. Clarkson, T. W., and S. R. Toole: Amer. J. Physiol. 206, 658 (1964).
11. Ussing, H. H., and B. Anderson: Proc. 3rd intern. Congress Biochem. Brussels, p. 434 (1955).
12. Cooperstein, I. L., and C. A. Hogben: J. gen. Physiol. 42, 461 (1959).

7*

13. Leaf, A., J. Anderson, and L. B. Page: J. gen. Physiol. **41**, 657 (1958).
14. Eigler, F. W., U. Held, W. Karger, and K. J. Ullrich: Pflügers Arch. ges. Physiol. **270**, 79 (1959).
15. Donn, A., and D. M. Maurice: Arch. Ophthal. **62**, 741, 748 (1959).
16. Cole, D. F.: Brit. J. Ophthal. **46**, 577 (1962).
17. Cole, K. S.: Arch. Sci. physiol. **3**, 253 (1949).
18. Marmont, G.: J. cell. comp. Physiol. **34**, 351 (1949).
19. Hodgkin, A. L., A. F. Huxley, and B. Katz: J. Physiol. (Lond.) **116**, 424 (1952).
20. Hays, R. M., and A. Leaf: J. gen. Physiol. **45**, 905, 921, 933 (1962).
21. Essig, A., and A. Leaf: J. gen. Physiol. **46**, 505 (1963).
22. Anderson, B., and K. Zerahn: Acta physiol. scand. **59**, 319 (1963).
23. Ussing, H. H., T. U. L. Bieber, and N. S. Bricker: J. gen. Physiol. **48**, 425 (1964).
24. Lindley, B. D., and T. Hoshiko: J. gen. Physiol. **47**, 773 (1964).
25. Finkelstein, A.: J. gen. Physiol. **47**, 545 (1964).
26. Ussing, H. H., and E. E. Windhager: Acta physiol. scand. **61**, 484 (1964).
27. Curran, P. F., and M. Cereijido: J. gen. Physiol. **48**, 543, 1011 (1965).
28. Ussing, H. H.: Acta physiol. scand. **63**, 141 (1965).
29. Bullivant, S., and W. R. Loewenstein: J. Cell Biol. **37**, 621 (1968).
30. Gatzy, J. T.: Fed. Proc. **25**, 507 (1966).
31. Tormey, J. McD., and J. M. Diamond: J. gen. Physiol. **50**, 2031 (1967).
32. Ussing, H. H.: J. Cell Biol. **36**, 625 (1968).
33. Lindemann, B.: Biochim. biophys. Acta (Amst.) **163**, 424 (1968).
34. Bueno, E. J., and L. C'orchs: J. gen. Physiol. **51**, 785 (1968).
35. Hays, R. M.: J. gen. Physiol. **51**, 385 (1968).

Questions for Discussion

1. What is meant by "steady-state"? Can the tissue be in "steady-state" with regard to net flow of sodium and not be in steady-state with regard to fluxes of ^{24}Na?

2. What evidence do we have to indicate when the tissue is in steady-state with regard to total sodium and isotopic sodium?

3. Does this experiment allow estimation of all the rate constants $(k_1 \ldots k_4)$ with equal accuracy? If not, which is determined with least error, which with most?

4. Antidiuretic hormone increases, Ca^{++} decreases, the permeability of the outer surface of the skin to sodium. How will these agents affect your experimental observations (i.e. M_{12}, M_{21}, M_{23}, M_{32}, S_2, SCC)? Ouabain, a cardiac glycoside, inhibits the sodium pump. What effect would this agent have on your observations?

5. In many experiments reported in the literature, biological membranes are regarded as a single membrane, not as a compartment as in this model of frog skin. The unidirectional fluxes reported are equivalent to M_{13} and M_{31} i.e. they are the unidirectional fluxes from compartments 1 to compartment 3. How do M_{13} and M_{31} relate to the individual fluxes

Sodium — 24

λ factor (10 min) = 0.00767 $T_{1/2}$ = 15.06 h

min h	0	10	20	30	40	50
0	—	.9923	.9848	.9773	.9698	.9624
1	.9550	.9477	.9405	.9333	.9262	.9191
2	.9121	.9051	.8982	.8913	.8845	.8777
3	.8710	.8644	.8578	.8512	.8447	.8383
4	.8319	.8255	.8192	.8129	.8067	.8005
5	.7944	.7884	.7823	.7764	.7704	.7645
6	.7587	.7529	.7471	.7414	.7358	.7302
7	.7246	.7190	.7136	.7081	.7027	.6973
8	.6920	.6867	.6815	.6762	.6711	.6659
9	.6609	.6558	.6508	.6458	.6409	.6360
10	.6311	.6263	.6215	.6167	.6121	.6074
11	.6027	.5981	.5936	.5890	.5845	.5801
12	.5756	.5712	.5668	.5625	.5582	.5539
13	.5497	.5455	.5413	.5372	.5331	.5290
14	.5250	.5210	.5170	.5130	.5092	.5053
15	.5014	.4975	.4938	.4900	.4862	.4825
16	.4788	.4752	.4715	.4679	.4644	.4608
17	.4573	.4538	.4503	.4469	.4435	.4401
18	.4367	.4334	.4301	.4268	.4235	.4203
19	.4171	.4139	.4107	.4076	.4045	.4014
20	.3983	.3953	.3922	.3893	.3863	.3833
21	.3804	.3775	.3746	.3717	.3689	.3661
22	.3633	.3605	.3578	.3550	.3523	.3496
23	.3469	.3443	.3417	.3391	.3365	.3339
24	.3313	.3288	.3263	.3238	.3213	.3189
25	.3164	.3140	.3116	.3092	.3069	.3045
26	.3022	.2999	.2976	.2953	.2931	.2908
27	.2886	.2864	.2842	.2820	.2799	.2777
28	.2756	.2735	.2714	.2694	.2673	.2652
29	.2632	.2612	.2592	.2572	.2553	.2533
30	.2514	.2495	.2476	.2457	.2438	.2419
31	.2401	.2382	.2364	.2346	.2328	.2310
32	.2293	.2275	.2258	.2241	.2224	.2206
33	.2190	.2173	.2156	.2140	.2124	.2107
34	.2091		.2059		.2028	
35	.1997		.1967		.1937	
36	.1907		.1878		.1850	
37	.1821		.1794		.1766	
38	.1739		.1713		.1687	
39	.1661		.1636		.1611	

Table 2. *Primary Data*

Chamber (Medium)	Sample No.	Sample Volume ml	Time of sampling h	min	Time of counting	Short-Circuit current chambers Time h:min	1 μA	2 μA
1_t	1	0.5	4	$1^1/_2$	The times were	4:00	136	108
1_t	2	0.5		3	taken from the	4:05	134	112
1_t	3	0.5		$4^1/_2$	print-out of the	4:10	132	116
1_t	4	0.5		6	automatic γ counter	4:13	130	118
1_t	5	0.5		$7^1/_2$		4:20	130	120
1_t	6	0.5		9		4:25	128	114
1_t	7	0.5		11		4:30	128	112
1_t	8	0.5		13		4:35	128	110
1_t	9	0.5		$14^1/_2$		4:40	124	105
1_t	10	0.5		16		4:45	122	102
1_t	11	0.5		$17^1/_2$		4:50	120	100
1_t	12	0.5	4	59		4:55	118	96
2_o	13	0.5	5	01		5:00	114	92
1_t	14	0.5	5	20		5:05	112	92
2_o	15	0.5	5	21		5:10	112	90
1_t	16	0.5	5	41		5:15	111	88
2_o	17	0.5	5	42		5:20	110	88
1_t	18	0.5	5	59		5:25	109	87
2_o	19	0.5	6	00		5:30	108	86
1_t	20	0.5	6	20		5:35	108	84
2_o	21	0.5	6	21		5:40	106	84
1_o	22	0.5 ml diluted to 100 ml, 0.5 ml counted	6	55		5:50	104	84
2_t	23	0.5 ml diluted to 100 ml, 0.5 ml counted	6	55		5:55	104	82
Slin 1	24					6:05	102	80
Skin 2	25					6:10	102	80
						6:15	101	80
						6:20	100	80

M_{12}, M_{21}, M_{23} and M_{31} across each membrane? What advantage arrives from measuring the individual fluxes across each membrane rather than the fluxes between the two bathing solutions?

Table 3. *Solution Volumes*

Solutions were removed at the end of the experiment and the volumes measured by weighing.

Solution	1_i	1_o	2_i	2_o
Final volume (ml)	9.5	9.8	9.0	9.8
Initial volume (ml)	10.0	10.0	10.0	10.0
Mean volume (ml)	9.75	9.9	9.5	9.9

Table 5. *Computations on Skin Counts*

Steps in calculation	(1)	(2)	(3)	(6)
Sample No.	Counts/min	Background corrected Counts/min	Decay corrected Counts/min	$P_{2\infty}$
24	35370	34763	41468	26954
25	438596	437990	526490	

Calculation of M_{13} and M_{31} and Comparison with the SCC

The calculations are based on the equations on page 91

$$M_{13} = \frac{0.117 \ (\text{Equiv}/l)}{6.157 \cdot 10^7 \ (\text{counts}/\text{min})} \cdot \dot{P}_{3\infty} \ (\text{counts}/\text{min}) \cdot 9.9 \times 10^{-3} \ (l) \ \text{equiv}./\text{min}$$

$$M_{31} = \frac{0.117 \ (\text{Equiv}/l)}{6.893 \cdot 10^7 \ (\text{counts}/\text{min})} \cdot \dot{P}_{1\infty} \ (\text{counts}/\text{min}) \cdot 9.5 \times 10^{-3} \ (l) \ \text{equiv}./\text{min}$$

$$\text{SCC} = \frac{\mu\text{A} \cdot 10^{-6} \ (\text{coul}/\text{sec})}{96{,}500 \ (\text{coul}/\text{equiv})} \cdot 60 \ (\text{sec}/\text{min}) \ \text{equiv}./\text{min}$$

Time	M_{13}	M_{31}	$M_{13} - M_{31}$	Step (14)
	10^{-8} equiv./min			SCC
5:00 – 5:20	7.14	0.76	6.38	6.96
5:20 – 5:40	9.15	0.79	8.34	6.72
5:40 – 6:00	5.55	0.84	4.71	6.49
6:00 – 6:20	6.80	0.96	5.84	6.25
Mean			6.30	6.60 (Step 13)

Table 4. *Computations Solution Counts*

Steps in calculations Sample No.	(1) Counts counts/min	(2) Background corrected counts/min	(3) Decay corrected counts/min	(4) Total counts in chamber counts/min	(5) Total counts in chamber corrected for sampling counts/min	(9) P_3 counts/min²	(9) $\dot P_3/\dot P_{3\infty}$	(9) $(1-\dot P_3/\dot P_{3\infty})$ counts/min²
1	218	7	7	136	136	90	.023	.977
2	253	42	43	838	845	473	.119	.881
3	353	142	144	2808	2858	1342	.339	.661
4	486	275	281	5480	5674	1877	.474	.526
5	684	473	488	9516	9991	2878	.727	.273
6	902	691	718	14000	14963	3315	.837	.163
7	1256	1045	1094	21332	23013	4025		
8	1551	1340	1414	27572	30347	3667		
9	1756	1545	1643	32038	36227	3920		
10	2055	1844	1976	38532	44364	5424		
11	2119	1908	2060	40190	47998	2422		
12	10900	10684	11624	226668	236536	$\dot P_{1\infty}$ a. $\dot P_{3\infty}$ (step 8) cts/min²		
13	1900	1281	1404	27798	27798			
14	14400	13827	15277	297902	319394	3946		
15	2230	1624	1808	35798	37202	470		
16	18400	17771	19938	388790	425559	5055		
17	2600	1976	2234	44234	47446	488		
18	19700	19088	21748	424086	480793	3069		
19	2870	2260	2595	51380	56826	521		
20	21940	21332	24678	491220	559675	3756		
21	3240	2636	3073	60846	69365	597		
22	14054	13541	15549	6.157×10^7				
23	15830	15317	18139	6.893×10^7				

Mean $\dot P_{1\infty}$ = 519

Mean $\dot P_{3\infty}$ = 3960

Calculation of Isotopic Rate Constants

From the slope of the graph of log $(1 - \dot{P_3}/\dot{P}_{3\infty})$ against time (see Step 9),

Step (9)　　　　$k = 0.300$ min^{-1}

Step (11)　　　　$k_{23} = \dot{P}_{3\infty}/P_{2\infty}$　$= 3960/26{,}954$ min^{-1}
　　　　　　　　　　　　　　　　　$= 0.147$ min^{-1}

Step (12)　　　　$k_{21} = k - k_{23}$　　$= 0.153$ min^{-1}

Step (7) + (10)　$k_{12} = k \dfrac{P_{2\infty}}{P_1} = .300 \times 26{,}954/6.157 \times 10^7$ min^{-1}
　　　　　　　　　　　　　　$= 0.131 \times 10^{-3}$ min^{-1}

Step (14) Mean short-circuit-current, Chamber 1 = 117 μA =
= 7.27×10^{-8} equiv. min^{-1}

Step (15)　　　　$S_1 = 0.117 \times 9.9 \times 10^{-3}$ equiv. Na$^+$
　　　　　　　　　　1.158×10^{-3} equiv. Na$^+$

Step (16)　　　　$S_2 = (k_{12} S_1 - \text{SCC}/F)/k_{21}$
　　　　　　　　　　$= (0.131 \times 10^{-3} \times 1.158 \times 10^{-3} - 7.27 \times 10^{-8})/$
　　　　　　　　　　0.153 equiv.
　　　　　　　　　　$= 5.16 \times 10^{-3}$ equiv.

Step (17)　　　　$S_3 = 0.117 \times 9.75 \times 10^{-3}$ equiv.
　　　　　　　　　　$= 1.14 \times 10^{-3}$ equiv. Na$^+$

Step (18)　　　　$k_{32} = (k_{23} S_2 - \text{SCC}/F) S_3$
　　　　　　　　　　$= (0.147 \times 5.16 \times 10^{-7} - 7.27 \times 10^{-8})/$
　　　　　　　　　　1.14×10^{-3} min^{-1}
　　　　　　　　　　$= 0.28 \times 10^{-5}$ min^{-1}

Comment. The error in the estimation of k_{32} is large. The chief reason is that two steps in the calculation, steps (16) and (18) involve the subtraction of two products having similar numerical values.

Calculation of Unidirectional Fluxes

Step (19)　　　　$M_{12} = k_{12} S_1 = 0.131 \times 10^{-3} \times 1.158 \times 10^{-3}$ equiv./min
　　　　　　　　　　　　$= 15.1 \times 10^{-8}$ equiv./min

　　　　　　　　$M_{21} = k_{21} S_2 = 0.153 \times 5.16 \times 10^{-7}$
　　　　　　　　　　　　$= 7.9 \times 10^{-8}$ eqiv./min

　　　　　　　　$M_{23} = k_{23} S_2 = 0.147 \times 5.16 \times 10^{-7}$
　　　　　　　　　　　　$= 7.6 \times 10^{-8}$ equiv./min

　　　　　　　　$M_{32} = k_{32} S_3 = 0.28 \times 10^{-5} \times 1.14 \times 10^{-3}$
　　　　　　　　　　　　$= 0.3 \times 10^{-8}$ equiv./min

Micropuncture and Microanalysis in Kidney Physiology

By K. J. Ullrich, E. Frömter, and K. Baumann

I. Introduction

Micropuncture techniques and microanalysis have been the prerequisites for the investigation of two major questions in kidney physiology:

1. At what sites along the nephron are the different substances transported into or out of the tubular urine;

2. How do these transport mechanisms work, e.g. by what forces and through which mediating structures are substances moved from one compartement into another and how are these processes regulated.

Sampling tubular urine under free flow conditions from different sites along the nephron and measuring changes in concentration permits localization of transport processes. This technique has been applied to study net fluxes across individual tubular segments for many substances. More elaborate techniques are required to study the second question and as a consequence our knowledge of the basic physico-chemical mechanisms underlying the different transport processes has not yet advanced as far.

The techniques to be described have been applied in our laboratories to determine the permeability of the tubular epithelium to different substances, to measure reflection coefficients and electrical parameters and thus to differentiate between active and passive transport rates.

II. Preparation of Animals and Immobilization of the Kidney

The animal is anesthetized by intraperitoneal injection of Inactin, [ethyl-(1-methyl-propyl)-thiobarbital sodium, Promonta GmbH, Hamburg, Germany]. A dose of 80 mg/kg body weight assures complete anesthesia, which lasts for several hours. The animal is kept right side down on a thermostatically-controlled heated table (Fig. 1 A). The rectal temperature is continuously monitored to prevent the animal from being overheated. After lateral incision of the abdominal wall the left kidney is exposed and mobilized gently by removing the perirenal fat. The thin fibrous capsule can be either removed or preserved. For sampling of urine and to avoid obstruction of urine flow the ureter is

canulated with a polyethylene catheter redrawn to an outer diameter of
$300 - 400 \mu$. A little cup [83] is inserted into the abdominal cavity
to hold and immobilize the kidney. Best immobilization has been obtained with a double cup [21] (Fig. 1 C and D). The outer cup prevents
dislocations from respiratory movements of the neighbouring organs
and the inner one holds the kidney. To further improve immobilization
the inner cup is filled with Agar (3% in Ringers solution), which stiffens
at 38 °C. The area to be punctured is left uncovered. To protect against

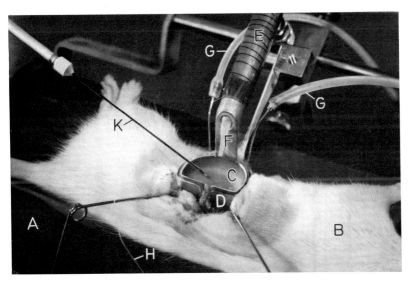

Fig. 1. Experimental set up. A Heated animal table. B Rat. C Inner kidney cup
(metal, coated with epoxy cement) containing the kidney which is embedded in
Agar and covered with Tyrode solution. D Outer kidney cup (Lucite). E Metal
sidearm holding inner kidney cup. It is wound with a heating wire. F Polyethylene
tubing supplying warm Tyrode solution to the kidney surface. G Polyethylene
tubing sucking off overflowing Tyrode solution from outer kidney cup. H Catheter
inserted into the ureter. K Micropipette, filled with black oil, inserted into a holder

drying, warm paraffin oil or Tyrode solution is continuously dropped onto
the exposed surface (Fig. 1 F).

The micropipettes are inserted into a holder (Fig. 1 K), which is
mounted on a micromanipulator (Leitz, Wetzlar, Germany). During puncture the kidney surface is illuminated by a Monla lamp (Leitz) with a heat
absorption filter and observed through a stereo microscope (Leitz) at 100
to $218 \times$ magnification. The room is darkened to improve visualization.

For the investigation of medullary function several techniques have been used
to gain access to the renal medulla. When the ureter is cut transversely, as high up

as possible, the papilla of the golden hamster or the young rat protrudes 2 to 3 mm and can readily be punctured. It is recommended to work with transillumination delivered via a bent quartz rod (Diaskleral-lamp Zeiss, Germany). Since the concentrating ability of the exposed papilla vanishes in this preparation slowly, attempts have been made to puncture through a small incision in the ureter. An entirely different way to expose the renal papilla was reported by Sakai et al. [57]. A part of the cortex is excised to gain access to the renal pelvis from the kidney surface. The papilla is brought up through this cortical window and fixed with acrylate adhesive. By this technique up to 5 mm of medullary tissue is made accessible to micropuncture.

III. Preparation of Micropipettes

1. Single Barreled Micropipettes. In our laboratories soft glass tubing of 0.9 mm o.d. and 0.5 mm i.d. (Scientific Glass Apparatus, Bloomfield N.J., USA) is used. The glass is cleaned, mounted in a horizontal spring operated pulling machine [10] and pulled to give two capillaries having a

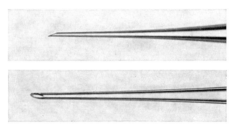

Fig. 2. Tip of micropipette after sharpening. Outer diameter 7 μ

relatively short shank. After pulling, the tip is cut with a pair of fine scissors to an o.d. of 6 to 10 μ under a stereoscopic microscope at 50 × magnification. It is then bevelled with a grinding machine. The machine has a motordriven rotating disk, which consists of an alloy of lead and tin and which is covered with a thin layer of diamond grinding paste (Winter-Diaplast N 0.25, alcohol- and watersoluble, available from Winter u. Sohn, Hamburg 19, Germany). The micropipette is clamped in a stand and lowered to the rotating surface at an angle of 30°. For beginners it is recommended to watch this procedure under the microscope at a magnification of 20 to 40 ×. Sharpening is effected in 1/2 to 1 min. The tip of the capillary is cleaned by dipping into sulfur-chromic acid, rinsing several times with distilled water, and removing the water by a jet of air. Finally, the sharpness and dimension of the tip are checked under the microscope at 400 × magnification (Fig. 2). In other laboratories micropipette sharpeners are used which have a grinding stone [27, 71].

2. Double-Barreled Micropipettes. Two glass capillaries are fixed together firmly at two points (Fig. 3 A) with sealing or dental wax. The middle portion is softened in a microflame (see below) and twisted around two or three times to melt both glass tubings together (Fig. 3 B). The twisted part is then predrawn by hand in the microflame to an outer diameter of around 0.3 to 0.5 mm (Fig. 3 C). Care must be taken to prevent bending. The final pulling is done in the machine described above. If capillaries break during fixation in the pulling machine due to axial deviation, fixation can be accomplished with Scotch tape.

3. Ling-Gerard-Microelectrodes. In our laboratories microelectrodes are drawn from Pyrex glass capillary tubing (Corning, New York) of 1.5 mm o.d. and 1.0 mm i.d. with an Alexander-Nastuk-type [2] vertical

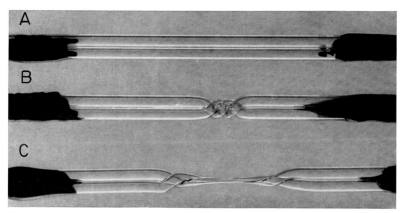

Fig. 3. Production of double-barreled micropipettes. The pipettes are shown after having been fixed together with sealing wax (A), twisted (B) and predrawn (C). Finally the tips are drawn out in a pulling machine

two-stage puller. It is important to clean the glass capillaries before pulling (detergent, aqua dest, acetone) and to handle them by forceps only. The electrodes are suspended in a lucite holder, tip down and filled according to TASAKI et al. [61] in methanol under negative pressure. The electrolyte solution (usually KCl, 3 mol/l) is allowed to diffuse into the capillaries over night. In order to avoid high tip potentials the solution must be filtered shortly before use (Ultrafeinfilter UFF fein, pore diameter $10 - 20 \times 10^{-9}$ m from Membranfilterges. Göttingen, Germany).

IV. Micropuncture Techniques

1. Identification of Proximal and Distal Tubules. Transit Time. Although it is possible with some practice to distinguish proximal from distal tubules by their microscopic appearance alone, the discrimination is much easier with the method described by STEINHAUSEN [60].

Fig. 4

Lissamine green, 0.03 ml of a 10% aqueous solution (Chroma-Gesellschaft Schmid & Co., Stuttgart-Untertürkheim, Postfach 146, Germany) is injected as rapidly as possible into the jugular vein. A few seconds later the kidney shows a "green flush", indicating the first dye transit through the blood capillaries (Fig. 4a). At this time the dye has also reached the glomeruli, which lie beneath the kidney surface. Immediately thereafter the filtered green dye appears in the first surface loops of the proximal tubules (Fig. 4b). The dye front travels through the proximal convolutions in about 9 sec (Fig. 4c and d). Thereafter, the whole dye column disappears (Fig. 4e) from the surface but reappears about 30 sec later in single strongly colored loops of the distal convolution (Fig. 4f – h). In addition to distinguishing between proximal and distal tubules the method allows us to identify first and last loops and to determine the transit time. Furthermore it gives an idea of whether the blood circulation of the kidney is uniform and adequate or not. The *transit time* of the proximal convolution is defined as the interval from the "flush" to the appearance of a dye front in the last loop of the proximal convolution.

2. Collection of Free Flow Samples of Tubular Urine [75]. Sharp micropipettes with an outer diameter of $5 - 8\,\mu$ are selected. They are filled with colored oil in order to avoid evaporation of samples within the pipette. Paraffin or castor oil is saturated with Sudanblack, filtered and centrifuged at 10 000 r.p.m. for 1 h to remove undissolved dye particles. The pipettes are filled from the large open end with a syringe via a length of polyethylene tubing.

When an ideally sharp pipette is used with a good manipulator, micropuncture per se does not require very much skill. The pipette is pushed through the tubular wall at an angle of 10° to 15°. After penetration into the lumen tubular fluid should enter the pipette spontaneously. If this is not the case the intraluminal position of the tip should be checked since the tip might touch the tubular wall. A small oil droplet may be injected into the tubular lumen to confirm its position. Sometimes it is necessary

---

Fig. 4. Kidney surface. Passage of Lissamin green through the proximal and distal convoluted tubules. a 3 sec after injection: the dye is visible in four postglomerular capillary areas marked by long arrows. b 9.5 sec after injection: the dye is visible in proximal loops of proximal convolutions. c and d 14 and 21 sec after injection: the dye has reached the last loops of the proximal convolutions. e color has disappeared from the surface. f to h 85, 95 and 123 sec after injection: distal tubules (indicated by long arrows) are heavily stained. (The short arrow marks the same point in all pictures). For technical reasons passage time has been artificially prolonged in this experiment. By courtesy of M. STEINHAUSEN [60]

to suck a small amount of fluid into the tip to start the spontaneous inflow caused by intratubular pressure.

In contrast to the tubular puncture itself the collection of true free flow samples requires some experience and should be done most carefully. If too much fluid is withdrawn from the tubule — either by application of negative pressure or simply by capillarity — the tubular flow will change. Pressure and transit time will decrease, while filtration rate increases. The analytical data from such samples are of little value. To avoid mistakes the following suggestions may be helpful:

1. Allow the fluid to enter very slowly; apply a little positive counter-pressure to the pipettes;

2. Observe the tubular diameter and prevent the tubule from collapsing;

3. Calculate from sampling time and sample size the fraction aspirated from the estimated total volume flow at the puncture site. It should not exceed 10 to 20%;

4. If possible measure the intratubular hydrostatic pressure before and during sampling and keep it constant.

When the size of the sample is adequate for analysis, the pipette is withdrawn from the tubule and a small oil droplet is sucked into the tip to guard against evaporation.

3. Determination of Glomerular Filtration Rate of Single Nephrons. This technique is very similar to free flow collection, except that distal to the sampling site the lumen is blocked with a long oil column while all fluid delivered proximal to this is collected [18]. It is obvious, that reliable measurements can only be obtained, when the intratubular pressure of the tubule punctured is controlled simultaneously and kept constant during sampling. Single nephron GFR is calculated from inulin ratio and fluid volume collected per unit time.

4. Localization of Puncture Site. After sampling has been finished the entire length of the nephron is filled with neoprene [6] or latex, and later the cast is dissected under the microscope. According to MALNIC et al. [46] latex solution is prepared by mixing 1 ml of latex (available from Turtox General Biological Supply House Inc. Chicago 20 Ill. USA) and 4 ml of Aquarex D solution (10 mg per ml H_2O) in a test tube under oil to hinder access of oxygen. A small amount of this stock solution is sucked into the tip of an oil-filled micropipette ($12\,\mu$ o.d.) and protected against air by an additional oil-droplet. The latex is injected slowly into the tubule under increasing pressure. Any interruption of injection will lead to blockage of the tip. Then the location of the punctured loop with respect to the other loops of the same nephron is protocolled in a sketch for better orientation during dissection. After maceration in 6 n HCl at 38 °C for 90 min the kidney is transferred to a water filled Petri dish and dissected under the stereomicroscope (magnification 12 − 100 ×) with a pair of needles. The latex cast is isolated from surrounding tissue; glomerulus, puncture site and end of the proximal tubule are verified and the parts between are cut into short segments, the length of which can be measured with an eye-piece micrometer.

5. Continuous Microperfusion of Individual Tubular Segments. This technique has been described in detail by SONNENBERG and DEETJEN [58]. The flow of tubular urine is blocked proximal to the puncture site with oil and the rest of the nephron is perfused continuously with a test solution at constant speed. Samples of the perfusate are collected from single loops reappearing at the kidney surface and changes in concentrations of the perfusate are determined (Fig. 5).

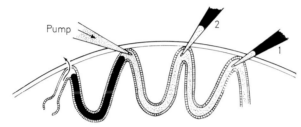

Fig. 5. Principle of continuous microperfusion. When collection with pipette Nr. 1 is finished the tubule is punctured more proximally with micropipette Nr. 2 and a second sample is obtained. Further details are described in the text

Often it is advantageous to perfuse the tubules with a steady state solution containing 110 mMol/l NaCl and 80 mMol/l Raffinose [36] in order to bring net water flux to zero. Under this condition solute fluxes across the tubular wall can readily be calculated [59] from perfusion rate, inner tubular diameter, concentration changes and lengths of perfused segments which are measured by microdissection of latex-casts. If transtubular water flux is not zero proper corrections have to be applied.

Application of the perfusion technique requires some practice. First a tubule is punctured with an oilfilled micropipette (A) and a small oil droplet is injected into the fluid stream to outline course and sequence of the loops. If there are at least three additional loops downstream, the pumping pipette (B) is inserted into the first of them. The segment in between is filled with oil, while the pump is running. Then pipette A is withdrawn to allow drainage of glomerular filtrate and is reinserted for sampling into the last accessible loop (Fig. 5, pipette 1). Additional samples can be obtained by puncturing more proximally (Fig. 5, pipette 2) [44].

8   Membrane Biophysics

Fig. 6 shows a diagram of the head of the microperfusion pump. The space between pipette holder (f) and teflon ring (g) (gasket) is filled with mineral oil and the pipetteholder is screwed to the pump body (h) tightly in order to compress the teflon ring (g) and to seal the stem (k) within this ring. Then the micropipette (a), filled with the test solution, is inserted into the pipette holder and sealed by nut (c) and gasket (e). During filling, care must be taken to avoid air bubbles. An adjustable electromotor advances stem (k) slowly in a screw-like motion thereby displacing volume from the sealed chamber formed by pipette holder (f) and pipette (a). Fluid can be delivered at a constant rate from 6 to 50 nl/min. The lucite tube (b) protects the pipette against air drafts which would change the temperature and disturb the constant

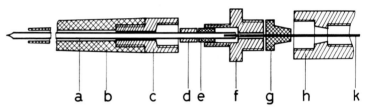

a   b   c   d e   f   g   h   k

Fig. 6. Microperfusion pump (detail). a Micropipette. b Lucite tube protecting micropipette from air drafts. c Nut. d Small metal tube. e Short piece of silicone tubing which is compressed by c and d and seals the pipette within the holder. f Pipette holder. g Teflon gasket, which is compressed by the pipette holder and seals the stem. h Body of the pump. k Stem, which advances by rotation, driven by an electromotor

delivery by volume expansion-contraction. The rate of fluid delivered by the microperfusion pump can be calibrated in two different ways:

1. A pipette is inserted into the pump and the movement of a fluid column within the pipette watched under the microscope. When the advancement of the column per unit time has been determined, the pipette is cut transversely, cleaned, fixed in an *upright* position under the microscope and the inner diameter of the glass tube is read with an eyepiece micrometer. From linear velocity and diameter the rate of pumping is calculated.

2. The pump is allowed to inject radioactively marked fluid into a vessel containing scintillation fluid for a given time, and the radioactivity is counted to obtain the volume flow rate of the pump.

6. Determination of Net Absorption of Fluid from Single Tubular Loops (Split Oil Droplet Technique). This method has been developed by Gertz [24]. The tubule is punctured with a double-barreled micropipette, one barrel of which contains the test solution, while the other one is filled with black castor oil. An oil column is injected into the tubular lumen and split into two parts by injection of a small droplet of test solution (see Fig. 7a). An additional injection of oil separates test solution from pipette and pushes the droplet to an undisturbed segment of the loop.

While the oil column placed distally from the test solution should be long, the one lying proximally should have a length of 200 to 300 μ. According to our experience this is long enough to prevent fluid from

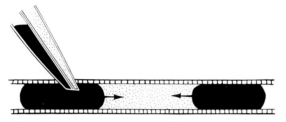

Fig. 7a. Split oil droplet technique. A column of black stained castor oil is injected into the tubule and split into two parts by injection of the test solution (dotted area). When test solution is absorbed by the tubular wall the oil columns approach each other

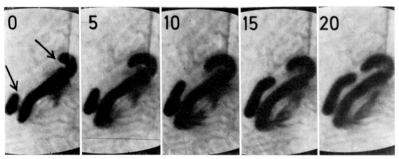

Fig. 7b. Series of photographs taken at 5 sec intervals after injection of test solution into a proximal tubule blocked with black castor oil. The gap within the oil column, indicated by the arrows, is formed by the test droplet and shrinks during resorption

entering or escaping between oil surface and brush border and is sufficiently short to allow the oil to move when the test droplet is shrinking. This is essential because the tubular diameter should be constant during resorption of the test droplet. It must be emphazised, that castor oil should be used, although it increases the luminal diameter from around 20 to 30 μ. Paraffin oil, which does not distend to the same degree, gives incomplete sealing and may damage the brush border [79, 41].

The fluid absorption is determined from the decreasing length of the test droplet with time. This can be registered by a series of photographs taken at 5 sec intervals (see Fig. 7b). An automatic camera (Robot, Düsseldorf, Germany) attached to one ocular of the stereomicroscope proved suitable. The kidney surface is illuminated by a flashlight apparatus (Zeiss) through the glass rod of a diaskleral lamp (Zeiss). A timer device (Robot) synchronizes camera shutter and flashlight.

7. Determination of Steady-State Concentrations. Since the tubular wall is impermeable to substances like raffinose or polyethylene glycol a droplet of isotonic raffinose solution will not disappear from the tubular lumen when injected between the oil columns. It rather expands with influx of solutes and water. With rising intraluminal solute concentrations, finally, a state of quasi-equilibrium is attained, when passive influx is compensated for by active outward transport. The concentration differences found under these conditions provide information about the active or passive nature of transport processes [36]. To measure this the droplet is re-collected after the steady state condition has been reached. By pooling several samples within the same pipette the amount of fluid collected can be augmented sufficiently to permit analysis.

8. Determination of the Hydrostatic Pressure of Tubular Fluid. A compensation principle introduced to renal physiology by Wirz [83] and Gottschalk [28] is used. A tubule is punctured with a sharp micropipette of about 10 μ o.d., which is filled with colored Ringer solution. With the end of the pipette open, tubular fluid would enter the tip and expel the colored solution. This can be prevented by application of counterpressure to the pipette. To obtain the intratubular pressure the reading has to be corrected for the capillarity of the pipette which can be measured in vitro or calculated from the dimensions of the tip. The method has been modified by Gertz to allow determination of hydrostatic pressure in glomerular capillaries [25].

9. Measurement of Electrical Potential Differences Across the Tubular Epithelium and Across Cell Membranes. Since electrophysiological techniques have been described extensively in the past [14, 16, 50] the following chapter will deal mainly with technical problems which have caused misunderstanding in the interpretation of electrical parameters of renal tubules of the rat [21].

For electrical measurements the kidney surface is conveniently covered with a continuously flowing bath of Tyrode solution approximately 1 mm deep. The microelectrode is inserted firmly into a compact lucite holder which is mounted on a micromanipulator. One sidearm of the holder contains a saturated calomel half cell through which electrical contact is established to the input of an electrometer. Through a second sidearm pressure of up to 10 atm can be applied to permit ejection of electrode fluid, if necessary. As indifferent electrode a sat. calomel halfcell is used which is connected to the kidney bath via a Ringer-agar-bridge.

The asymmetry of the circuit, which should be minimal under the conditions stated, is controlled before and after the experiment. This is done by insertion of a microelectrode with cut off tip into the kidney bath. Any difference between this value and the reading with an intact microelectrode is caused by its tip potential [1]. It is due to negative surface charges on the glass wall, which slow down chloride diffusion along the concentration gradient from capillary lumen to outside solution. Tip potentials are very unstable. They can even undergo changes during puncture, which are reversed when the electrode is drawn back into the bathing solution [5]. When the electrolyte solution surrounding the tip is diluted, the tip potential increases in proportion to its initial magnitude. This situation occurs during measurements in distal tubules. Therefore, care must be taken to select only microelectrodes with tip potentials of less than 2 mV.

Another criterion for the selection of good electrodes is their electrical resistance. This depends on the conductivity of filling solution and immersion medium, on the diameter of the tip, and to a smaller degree on its shape [19]. Knowledge of the resistance can provide some information about the size of the tip, which is invisible under the light microscope. We have successfully used KCl-filled electrodes (3 mol/l) of 1.5 to 100 MΩ for transepithelial measurements in both proximal and distal tubules [21]. Measurements of single membrane potentials, with the tip situated within a tubular cell, however, could only be performed with electrodes of 40 to 100 MΩ [78]. In order to exclude artifacts due to reversible blockage of the electrode tip during the puncture, a continuous check of the circuit resistance is recommended [30].

According to our experience the best way to improve visualization of the electrode in the bathing fluid is to work in a dark room with the light (Leitz Monla-lamp) at a very flat angle, producing a bright reflex on the glass wall. Attempts to fill the capillaries with colored solutions or to use fluorescence [52] have been discouraging.

For puncture, the microelectrode is lowered onto the kidney surface until it touches a tubule at an angle of around 10° to 15°. Then the tubule is impaled by a slight axial forward movement of the electrode. Best results are obtained, when the electrode enters the tubule along the tubular axis. To be certain of the exact intraluminal position of the electrode tip the following localizing methods are recommended:

1. Ejection of fluid from the electrode tip under visual control [26, 77, 12, 21].

2. Recording of potential profile [5] or better, simultaneous recording of potential steps and resistance steps during penetration of the tubular

wall with the microelectrode [45, 23]. This procedure requires two microelectrodes, one for the injection of current pulses, the other for measuring potential differences.

3. Observation of potential differences during changes of the ionic environment [23]. The single cellular membrane potential differences and the overall transepithelial potential difference respond differently to changes in ionic composition. Test solutions differing from plasma preferentially with respect to the concentration of Na and K are injected either into the tubular lumen or into the blood capillaries surrounding the tubule, and changes in potential difference are recorded and compared.

4. Iontophorectic deposition of dyes through the electrode tip into the tissue [19]. This method is time-consuming because the tissue must be prepared histologically and the position of the spot has to be verified under the microscope. It is used frequently by neurophysiologists and has been applied to kidney tubules of Necturus by Whittembury [76]. To our knowledge experience with this technique on rat tubules is lacking.

Another method was reported by Eigler [15]. A rise of circuit resistance during injection of oil into the impaled tubule was thought to indicate the intraluminal position of the electrode tip. This technique has also been applied to the rat kidney [33, 37, 48]. In our laboratory, however, it proved to give misleading results [30].

10. Determination of Specific Transepithelial Resistance. In order to determine the electrical resistance of the tubular wall cable analysis [50] has to be applied. In a first approximation the tubule can be regarded as a core conductor, the NaCl-filled lumen being the core, the epithelial wall representing the insulation. This simple model has indeed been shown to hold for the description of the overall time-independent potential change across the epithelium when current is applied through a point source intratubularly. Thus, the length constant derived from the observed voltage attenuation [82, 31] and the effective resistance measured [31] were found to fit the cable equations, when reasonable assumptions regarding the tubular diameter were made.

While the measurement of voltage drop along the tubule is relatively simple and the values reported appear to be safe many problems arise with the determination of the effective resistance (= input resistance). Any circuit designed to measure it with an electrode resistance in series is subject to severe errors, because of the instability of the electrode resistance. This often rises during puncture, reversibly or irreversibly, probably due to protein or membrane material which blocks the tip. Moreover it changes with changing conductivity of the immersion medium and is affected by hydrostatic pressure gradients across the tip. During application of negative pressure of 10 to 20 cm H_2O to the electrode — corresponding to the normal pressure difference across the tubular wall — we have seen electrode resistance increasing by several $M\Omega$. Double barreled electrodes on the other hand do not offer a real advantage, because their coupling resistance, which would lie in series with the effective resistance of the tubule, is still five times larger than the effective resistance and it, too, is unstable.

Another difficulty contributing to the uncertainty in determinations of the effective resistance is a deviation from pure exponential voltage decay in the very neighbourhood of the current point source (origin distortion) [17]. In view of these difficulties it seems advisable to use two separate microelectrodes for current injection and potential-measurements and to extrapolate eventually to the point of current injection from values measured at some distance [40].

A more detailed analysis, including single cell membrane resistance and cell to cell coupling has been attempted by WINDHAGER et al. [81] in Necturus kidney. The present day techniques, however, seem to be incapable of application in the rat kidney.

11. Perfusion of Blood Capillaries. Selective control of peritubular environment is possible by perfusing the peritubular capillaries with test solutions. This has been observed first with electrical measurements by FRÖMTER [22]. Recently this technique has been combined with the split oil droplet technique [68, 80] and with microperfusion [3].

After removal of the renal capsule a sharp micropipette of $7\,\mu$ o.d. is inserted into a blood capillary and test solution is injected by applying several atmospheres of pressure. The area which can be washed free of blood depends on the size of the blood capillary punctured and on the amount of pressure applied. Usually it consists of several tubular loops around the point of injection. Best perfusion is obtained when the venae stellatae are punctured.

V. Micro-Analysis

1. Sample-Size [32]. In a cortical nephron of the rat kidney the rate of flow of tubular fluid through the luminal cross section is as follows: In the glomerular capsule $\sim 20 \times 10^{-3}$, at the end of the proximal convolution $\sim 7 \cdot 10^{-3}$ and near the midpoint of the distal convolution $\sim 2 \times 10^{-3}\,\mu l/$ min. The luminal capacity (volume) of the non-dilated proximal convolution is $1.9 \times 10^{-3}\,\mu l$ (tubular radius $10\,\mu$ and length ca. $6000\,\mu$) and of the distal convolution $0.4 \times 10^{-3}\,\mu l$ ($8\,\mu$ radius and $2000\,\mu$ length). The sample volume available for microanalysis depends on localization, sampling time and sample mode, i.e. whether all tubular fluid reaching the sampling point is withdrawn or only a few percent of it. In free flow experiments usually $10 - 100 \times 10^{-3}\,\mu l$ are sampled, in stop flow perfusion experiments $0.2 - 1.5 \times 10^{-3}$ µl with each injection. The volume of such samples can be augmented by pooling.

2. Determination of Sample Volume. With most analysing procedures the sample is first blown out from the micropipette into the cavity of an oil-filled slide from which it may be reaspirated partially or totally for further handling.

In order to avoid excessive adhesion of sample fluid to glass, slides and measuring pipettes are siliconized. After thorough cleaning with water, chloroform, and acetone and after drying, the glassware is immersed in a solution of silicone oil in toluene $(5 - 10\% \text{ w/w})$ (Dow Corning silicone oil, $200 - 350$ cSt, available from Wacker Chemie, München, Germany) using a pair of forceps. Siliconization is completed by drying in an oven at 200 °C for 4 h. Alternatively Siliclad (Clay Adams Inc. New York, USA) diluted 100 times with water can be used. The clean glassware is dipped into this solution for 5 sec rinsed with water and dried either at room temperature for 24 h, or at 100 °C for 10 min. Before use, newly siliconized slides have to be washed with a brush several times to remove the silicon layer partially since otherwise the sample would not adhere to the glass at all. Small samples should be kept in the micropipette until beginning of analysis, since the surface area in contact with oil is rather small within the pipette and water loss from sample into the oil is minimal. This precaution makes use of water-saturated oil [4] dispensable.

In order to determine unknown concentrations it is sufficient to deliver identical amounts of sample and of standards to the reaction set-up, the exact amount of volume transferred being without relevance. Hence measuring pipettes need not be calibrated although this can be done when desired.

The filling of standard volume pipettes is done under a stereo microscope at 12.5 to 25 × magnification.

For small samples (10^{-4} to 10^{-2} µl) an oil-filled glass capillary of 1 mm o.d. is used, which is drawn out to a long thin tip and covered with a mark near the tip (Fig. 8 A). The mark is made either from nail lacquer or from a thin filament of dentistry-wax, which is attached to the tip by approaching a glowing glass rod.

Samples of 10^{-2} to 10^{-1} µl can be measured with capillaries of constant inner diameter (> 10 µ). They are sealed into bigger glass tubes for better handling (Fig. 8 B—D). The volume is marked with a filament of wax or the length of the fluid column is measured with an eyepiece-micrometer. When the sample has to be transfered it is recommended to seal the end of the measuring pipette with a small oildroplet in order to prevent evaporation.

When the entire length of the capillary is taken to measure the volume (Fig. 8 C) oil has to be prevented from entering the capillary. This is achieved by constantly blowing air through the tip, when the capillary is traversing the oil to reach the sample. To maintain good visualization of the sample the capillary is introduced from the side. Air bubbles are smaller when the ultimate tip is narrowed in a glowing platinum loop.

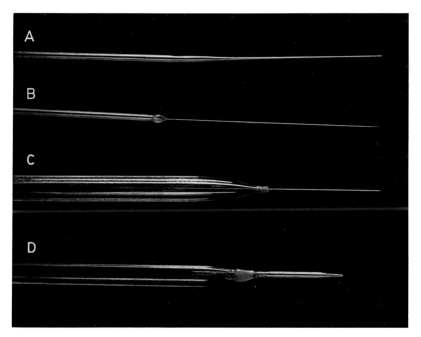

Fig. 8. Filling and standard volume pipettes, for details see text

Alternatively the capillary can be filled with chloroform, dipped into the oil, directed to the sample and emptied of chloroform just prior to insertion into the droplet.

3. Determination of Concentration.

a) Radioactivity measurements. Since mostly ³H or ¹⁴C labelled compounds are used, *liquid scintillation* counting is the method of choice. In this case the sample is blown directly into the solvent (0.2 ml Hyamine or NCS = solvent of Nuclear Chicago), and 15 ml scintillation fluid (1000 ml dioxan, 100 g naphthalene, 10 g PPO and 0,25 g POPOP) are added. Recently we have used a commercial PPO-POPOP mixture (Spectrafluor, Nuclear Chicago), which is diluted 1 : 25 with toluene.

b) Chemical analysis. *Bloodplasma* and *final urine* are usually available in amounts of more than 2 µl, which makes up with the addition of reagents to a final volume of about 200 µl. For samples in this order of magnitude test units and small polyethylene tubes are available on the market (Beckman-Spinco, Ultramicro Analytical System, München, Germany; Microlitersystem Eppendorf, Fa. Netheler u. Hinz, Hamburg, Germany). Since instruction for microanalysis in this dimensions are easily available and since almost all manufactures of photometers supply microcuvettes of suitable size, we confine our description here to a simple, convenient method for deproteination [70].

8 µl of water are placed on a piece of parafilm, 2 µl of sample (plasma, urine or standard) are pipetted into the droplet of water, and 2 µl of 10% zinc sulphate solution and 2 µl of 0.5 N sodium hydroxide are placed on the parafilm next to the water droplet. The three droplets are then mixed with a glass rod and drawn into a glass capillary 1.2 mm in diameter. The ends of the capillary are fused in a microflame and the capillary centrifuged. In spite of the evaporation that takes place during the pipetting and mixing on the parafilm, the method is extremely accurate. The explanation for this is that with some practice the time for each precipitation is so constant that all samples are exposed to essentially the same amount of evaporation.

The *micropuncture samples*, however, are so small, that even with addition of the reagents only 5 − 10 µl volume are available for photometry. For such small amounts a *microcuvette* was constructed by Ullrich and Hampel [64] Fig. 9 shows a schematic drawing of this microcuvette. The main part is a platinum tube, closed tightly on both sides with a piece of glass and a little sheet of polyethylene. Through two nozzles, inflow and outlet, the cuvette can be filled. The capacity of the cuvette is 2 − 5 µl so that 5 − 10 µl total volume for cleaning and filling is sufficient. The lightpath within the cuvette is 6 mm.

Commercially available photometers (Beckman or Zeiss) can be adapted to the microcuvette by addition of a special holder, which fixes the cuvette tightly in a constant position within the light path.

The reader is referred to the following publications for more detailed instructions regarding photometric analysis: Inulin [34, 73, 84], urea [39, 47, 69, 11], ammonia [65, 51], D-glucose [56, 44a, 84], lactic acid [56], protein [66], haemoglobin [67], calcium [20], paraaminohippuric acid [13], sulfonamides [59].

Several of the quoted methods require proper sealing of filled test tubes. This is done with a small flame (less than 5 mm in height) to avoid heating the nearby test solution. The flame burns on the drawn out tip (0.3 mm i.d.) of a glass tube and is fed by butane, propane or city gas.

c) By physical techniques. The apparatus and the procedure for the determination of *chloride* by electrotitration [55] for the measurement of *freezing point depression* with samples of only 10^{-3} µl [54, 35, 53] as well as for *microflamephotometry* [49, 46, 72] are explicitely described in the literature cited. All three methods are in routine use in our laboratory.

To measure the *pH value* of the microsamples we have repeatedly used the quinhydrone method [63, 51]. Two years ago KHURI and OELERT, working in our laboratory, built a glass electrode for sample size of

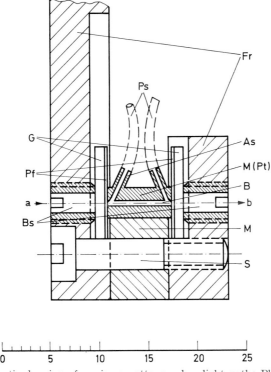

Fig. 9. Schematic drawing of a microcuvette. a – b = light path; Platinum tube [M (Pt) incl. B] with nozzles (As); glass windows (G) and sheets of polyetheylene (Pf) for tightening. BS and S screws, Fr and M brass housing, Ps filling tubes from polyethylene; scale at bottom in mm

10^{-1} µl [38]. Later UHLICH and BALDAMUS [62] improved this electrode by melting together the pH sensitive glass (titanglass and TH-glass A 41, Ingold) and the glass of the holder (leadglass 123 A, Osram). With this procedure electrical insulation is much better than that achieved by the use of agglutinants or cements. Fig. 10 gives a schematic diagram of the pH-microelectrode. The pH sensitive glass capillary (c) (approximate dimensions: i.d. 35 µ, o.d. 50 µ, length 2 mm) is connected by a glass-to-glass seal at point d to the inner capillary of lead glass (e) and at point b to the outer capillary of lead glass (f). The chamber between the inner and outer glass tubing is filled with reference solution (g, for example 0.1 n HCl) and covered with an oil droplet (k). After the insertion of a silver-silverchloride wire (h), which serves as reference electrode, the chamber is closed with epoxy cement (l). The whole assembly is then fixed in a thick glass tube (m) for better handling. Samples are sucked in through the tip (a) of the lead glass pipette (f) For micropuncture this pipette can be cut at an o.d. of 10 μ and sharpened in the grinding machine. While measuring the electrode remains

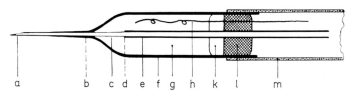

Fig. 10. pH-microelectrode for sample volumes of 0.01 μl as described in the text

immersed in the fluid sample to which electrical contact is established by means of a calomel half cell via a saturated KCl-bridge or by means of a Ling-Gerard-microelectrode of low resistance.

VI. Comments on Other Techniques

Finally some techniques should be mentioned, which have *not* been used in our laboratories. Some of them appear to be rather simple in performance but do not permit a quantitative evaluation of physico-chemical parameters of individual tubular segments. This is true for the technique of injecting radioactive substances into the tubular fluid or into the blood capillaries at various sites and measuring their appearance in the final urine of the experimental or control kidney [42]. Another technique which has been used in several laboratories is the determination of occlusion time, as described by Leyssac [43]. The renal artery is clamped and the time is measured, until the lumina of the proximal tubules visible at the kidney surface are completely collapsed. It had been thought that it might be possible to calculate the rate of net fluid resorption through the wall of the proximal tubules from that time or from the rate of change of luminal diameter. This, however, is not true, since during arterial clamping back flux of tubular urine into proximal surface tubules occurs [7, 74]. For many years kidney physiologists have dreamed of using isolated renal tubules like other isolated tissues e.g. frog skin and toad bladder. This dream has now become a reality after Burg et al. [9] reported their technique of isolating tubules of rabbit kidney and perfusing them in vitro. The results described by

GRANTHAM and ORLOFF [29] on collecting ducts and by BURG [8] on flounder tubules look very promising.

References

1. ADRIAN, R. H.: The effect of internal and external potassium concentration on the membrane potential of frog muscle. J. Physiol. (Lond.) **133**, 631 (1956).
2. ALEXANDER, J. T., and W. L. NASTUK: An instrument for the production of microelectrodes used in electrophysiological studies. Rev. Sci. Instrum. **24**, 528 (1953).
3. BAUMANN, K.: Unpublished observations.
4. BLOOMER, A. H., F. C. RECTOR JR., and D. W. SELDIN: The mechanism of potassium reabsorption in the proximal tubule of the rat. J. clin. Invest. **42**, 277 (1963).
5. BOROFFKA, I.: Elektrolyt-Transport im Nephridium von Lumbricus Terrestris. Z. vergl. Physiol. **51**, 25 (1965).
6. BOTT, P. A.: Renal excretion of creatinine in necturus. A reinvestigation by direct analysis of glomerular and tubular fluid for creatinine and inulin. Amer. J. Physiol. **168**, 107 (1952).
7. BRANDIS, M., G. BRAUN-SCHUBERT, and K. H. GERTZ: Retrograde flow of fluid within superficial tubules after blockade of glomerular filtration by aortic clamping. V. Symposium der Gesellschaft für Nephrologie, Lausanne 1967 (in press).
8. BURG, M.: Concentration steps in diodrast transport by flounder proximal tubules. Fed. Proc. **26**, 266 (1967).
9. —, J. GRANTHAM, M. ABRAMOW, and J. ORLOFF: Preparation and study of fragments of single rabbit nephrons. Amer. J. Physiol. **210**, 1293 (1966).
10. BURKHARDT, D.: Ultramikroelektroden aus Glas. Herstellung und Verwendung bei elektrophysiologischen Messungen. Glas und Instrumententechnik **3**, 115 (1959).
11. CAPEK, K., G. FUCHS, G. RUMRICH und K. J. ULLRICH: Harnstoffpermeabilität der corticalen Tubulusabschnitte von Ratten in Antidiurese und Wasserdiurese. Pflügers Arch. ges. Physiol. **290**, 237 (1966).
12. CLAPP, J. R., F. C. RECTOR, and D. W. SELDIN: Effect of unreabsorbed anions on proximal and distal transtubular potentials in the rat. Amer. J. Physiol. **202**, 781 (1962).
13. DEETJEN, P., u. H. SONNENBERG: Der tubuläre Transport von p-Aminohippursäure. Mikroperfusionsversuche am Einzelnephron der Rattenniere in situ. Pflügers Arch. ges. Physiol. **285**, 35 (1965).
14. DONALDSON, P. E. K.: Electronic apparatus for biological research. London: Butterworth 1958.
15. EIGLER, F. W.: Short-circuit current measurements in proximal tubule of necturus kidney. Amer. J. Physiol. **201**, 157 (1961).
16. FATT, P.: Intracellular microelectrodes. In: Methods in Medical Research, Vol. 9, p. 381. Chicago: J. H. Quastel (Editor-in Chief).
17. —, and B. KATZ: The electrical properties of crustacean muscle fibers. J. Physiol. (Lond.) **120**, 171 (1953).
18. FLANIGAN, W. J., and D. E. OKEN Renal micropuncture study of the development of anuria in the rat with mercury-induced acute renal failure. J. clin. Invest. **44**, 449 (1965).
19. FRANK, K., and M. C. BECKER: Microelectrodes for recording and stimulation, In: W. L. NASTUK Ed., Physical techniques in biological research, Vol. V. Part A, p. 23. New York and London: Academic press 1964.

20. FRICK, A., G. RUMRICH, K. J. ULLRICH, and W. E. LASSITER: Microperfusion study of calcium transport in the proximal tubule of the rat kidney. Pflügers Arch. ges. Physiol. **286**, 109 (1965).

21. FRÖMTER, E., u. U. HEGEL: Transtubuläre Potentialdifferenzen an proximalen und distalen Tubuli der Rattenniere. Pflügers Arch. ges. Physiol. **291**, 107 (1966).

22. —, C. W. MÜLLER und H. KNAUF: Fixe negative Wandladungen im proximalen Konvolut der Rattenniere uud ihre Beeinflussung durch Calciumionen. VI. Symposium der Gesellschaft für Nephrologie, Wien 1968 (in press).

23. —, T. WICK und U. HEGEL: Untersuchungen zur Ausspritzmethode für die Lokalisation der Mikroelektrodenspitze bei transtubulären Potentialmessungen an der Ratte. Pflügers Arch. ges. Physiol. **294**, 265 (1967).

24. GERTZ, K. H.: Transtubuläre Natriumchloridflüsse und Permeabilität für Nichtelektrolyte im proximalen und distalen Konvolut der Rattenniere. Pflügers Arch. ges. Physiol. **276**, 336 (1962).

25. —, J. A. MANGOS, G. BRAUN, and H. D. PAGEL: Pressure in the glomerular capillaries of the rat kidney and its relation to arterial blood pressure. Pflügers Arch. ges. Physiol. **288**, 369 (1966).

26. GIEBISCH, G.: Electrical potential measurements on single nephrons of necturus. J. cell. comp. Physiol. **51**, 221 (1958).

27. GLABMAN, S., R. M. KLOSE, and G. GIEBISCH: Micropuncture study of ammonia excretion in the rat. Amer. J. Physiol. **205**, 127 (1963).

28. GOTTSCHALK, C. W., and M. MYLLE: Micropuncture study of pressures in proximal tubules and peritubular capillaries of the rat kidney and their relation to ureteral and renal venous pressure. Amer. J. Physiol. **185**, 430 (1956).

29. GRANTHAM, J., and J. ORLOFF: Mechanism of potassium secretion in isolated perfused collecting tubules. Fed. Proc. **26**, 375 (1967).

30. HEGEL, U., u. E. FRÖMTER: Erfahrungen mit der Öltropfenmethode zur Lokalisation der Mikroelektrodenspitze bei transtubulären Potentialmessungen an der Rattenniere. Pflügers Arch. ges. Physiol. **291**, 121 (1966).

31. — — und T. WICK: Der elektrische Wandwiderstand des proximalen Konvolutes der Rattenniere. Pflügers Arch. ges. Physiol. **294**, 274 (1967).

32. HIERHOLZER, K., and K. J. ULLRICH: Grundzüge der Nierenphysiologie. In: Handb. Pharmakol. (In press.)

33. —, M. WIEDERHOLT, H. HOLZGREVE, G. GIEBISCH, R. M. KLOSE, and E. E. WINDHAGER: Micropuncture study of renal transtubular concentration gradients of sodium and potassium in adrenalectomized rats. Pflügers Arch. ges. Physiol. **285**, 193 (1965).

34. HILGER, H. H., J. D. KLÜMPER und K. J. ULLRICH: Wasserrückresorption und Ionentransport durch die Sammelrohrzellen der Säugetierniere. Pflügers Arch. ges. Physiol. **267**, 218 (1958).

35. KALBERT, F.: Direct reading biological cryostat. Clifton technical physics, Wanamassa N. J. USA.

36. KASHGARIAN, M., H. STÖCKLE, C. W. GOTTSCHALK, K. J. ULLRICH, and G. RUMRICH: Transtubular electrochemical potentials of sodium and chloride in proximal and distal renal tubules of rats during antidiuresis and water diuresis. Pflügers Arch. ges. Physiol. **277**, 89 (1963).

37. —, Y. WARREN, and H. LEVITIN (with the technical assistance of M. GARZASTY): Micropuncture study of proximal renal tubular chloride transport durin hypercapnea in the rat. Amer. J. Physiol. **209**, 655 (1965).

38. KHURI, R. M., S. K. AGULIAN, H. OELERT, and R. I. HARIK: A single unit pH glass ultramicroelectrode. Pflügers Arch. ges. Physiol. **294**, 291 (1967).

39. KLÜMPER, J. D., K. J. ULLRICH und H. H. HILGER: Das Verhalten des Harnstoffs in den Sammelrohren der Säugetierniere. Pflügers Arch. ges. Physiol. **267**, 238 (1958).

40. KNAUF, H., u. E. FRÖMTER: Messung des Kurzschlußstroms an den Speicheldrüsengängen des Menschen. Pflügers Arch. ges. Physiol. (In preparation.)

41. LANGER, K. H., W. THOENES und M. WIEDERHOLT: Licht- und elektronenmikroskopische Untersuchungen am proximalen Tubuluskonvolut der Rattenniere nach intraluminaler Ölinjektion. Pflügers Arch. **302**, 149 (1968).

42. LECHENE, C., et F. MOREL: Microinjections de sodium et d'inuline marqués dans les capillaires du rein de hamster. I. Perméabilité au sodium des segments tubulaires corticaux. Nephron **2**, 207 (1964).

43. LEYSSAC, P. P.: Dependence of glomerular filtration rate on proximal tubular reabsorption of salt. Acta physiol. scand. **58**, 236 (1963).

44. LOESCHKE, K., u. K. BAUMANN: Kinetische Studien der D-Glukoseresorption im proximalen Konvolut der Rattenniere. Pflügers Arch. **305**, 139 (1969).

44a. — —, H. RENSCHLER und K. J. ULLRICH: Differenzierung zwischen aktiver und passiver Komponente des D-Glukosetransports am proximalen Konvolut der Rattenniere. Pflügers Arch. ges. Physiol. **305**, 118 (1969).

45. LOEWENSTEIN, W., and Y. KANNO: Some electrical properties of a nuclear membrane examined with a microelectrode. J. gen. Physiol. **46**, 1123 (1963).

46. MALNIC, G., R. M. KLOSE, and G. GIEBISCH: Micropuncture study of renal potassium excretion in the rat. Amer. J. Physiol. **206**, 674 (1964).

47. MARSH, D. J., CH. FRAZIER, and J. DECTOR: Measurement of urea concentration in nanoliter specimens of renal tubular fluid and capillary blood. Analyt. Biochem. **11**, 73 (1965).

48. —, and S. SOLOMON: Relationship of electrical potential differences to net ion fluxes in rat proximal tubules. Nature (Lond.) **201**, 714 (1964).

49. MÜLLER, P.: Experiments on current flow and ionic movements in single myelinated nerve fibers. Exp. Cell. Res. Suppl. **5**, 118 (1958).

50. NASTUK, W. L., Ed.: Physical techniques in biological research, Vol. V. Electrophysiological methods. New York and London: Academic Press 1964.

51. OELERT, H., E. UHLICH und A. G. HILLS: Messungen des Ammoniakdruckes in den cortikalen Tubuli der Rattenniere. Pflügers Arch. ges. Physiol. **300**, 35 (1968).

52. PILLAT, B., u. P. HEISTRACHER: Ein einfaches Verfahren zur Sichtbarmachung von Glasmikroelektroden mit Hilfe von Fluorescin. Experientia (Basel) **16**, 519 (1960).

53. PRAGER, D. J., and R. L. BOWMAN: Freezing point depression. New method for measuring ultramicro quantities of fluid. Science **142**, 237 (1963).

54. RAMSAY, J. A., and R. H. J. BROWN: Simplified apparatus and procedure for freezing-point determination upon small volumes of fluid. J. sci. Instrum. **32**, 372 (1955).

55. RAMSAY, J. A., R. M. J. BROWN, and P. C. CROGHAN: Electrometric titration of chloride in small volumes. J. exp. Biol. **32**, 822 (1955).

56. RUIZ-GUINAZU, A., G. PEHLING, G. RUMRICH und K. J. ULLRICH: Glukose- und Milchsäurekonzentration an der Spitze des vaskulären Gegenstromsystems im Nierenmark. Pflügers Arch. Ges. Physiol. **274**, 311 (1961).

57. SAKAI, F., R. L. JAMISON, and R. W. BERLINER: A method for exposing the rat renal medulla in vivo: micropuncture of the collecting duct. Amer. J. Physiol. **209**, 663 (1965).

58. Sonnenberg, H., P. Deetjen und W. Hampel: Methode zur Durchströmung einzelner Nephronabschnitte. Pflügers Arch. ges. Physiol. **278**, 669 (1964).
59. —, H. Oelert, and K. Baumann: Proximal tubular reabsorption of some organic acids in the rat kidney in vivo. Pflügers Arch. ges. Physiol. **286**, 171 (1965).
60. Steinhausen, M.: Eine Methode zur Differenzierung proximaler und distaler Tubuli der Nierenrinde von Ratten in vivo und ihre Anwendung zur Bestimmung tubulärer Strömungsgeschwindigkeiten. Pflügers Arch. ges. Physiol. **277**, 23 (1963).
61. Tasaki, I., E. H. Polley, and F. Orrego: Action potentials from individual elements in cat geniculate and striate cortex. J. Neurophysiol. **17**, 454 (1954).
62. Uhlich, E., C. F. Baldamus und K. J. Ullrich: CO_2-Druck und Bicarbonatkonzentration im Gegenstromsystem der Nierenpapille. Pflügers Arch. ges. Physiol. **303**, 31 (1968).
63. Ullrich, K. J., F. W. Eigler und G. Pehling: Sekretion von Wasserstoffionen in den Sammelrohren der Säugetierniere. Pflügers Arch. ges. Physiol. **267**, 491 (1958).
64. —, u. A. Hampel: Eine einfache Mikroküvette für Monochromator Zeiss und Beckman Modell DU. Pflügers Arch. ges. Physiol. **268**, 177 (1958).
65. —, H. H. Hilger und D. J. Klümper: Sekretion von Ammoniumionen in den Sammelrohren der Säugetierniere. Pflügers Arch. ges. Physiol. **267**, 244 (1958).
66. —, G. Pehling und M. Espinar-Lafuente: Wasser- und Elektrolytfluß im vaskulären Gegenstromsystem des Nierenmarks. Pflügers Arch. ges. Physiol. **273**, 562 (1961).
67. — — und H. Stöckle: Hämoglobinkonzentration, Erythrocytenzahl und Hämatokrit im vasa recta Blut. Pflügers Arch. ges. Physiol. **273**, 573 (1961).
68. —, and G. Rumrich: The minimum requirements for the maintenance of sodium chloride reabsorption in the proximal convolution of mammalian kidney. J. Physiol. (Lond.) **197**, 69—70 P (1968).
69. — —, and B. Schmidt-Nielsen: Urea transport in the collecting ducts of rats on normal and low protein diet. Pflügers Arch. ges. Physiol. **295**, 147 (1967).
70. —, B. Schmidt-Nielsen, R. O'Dell, G. Pehling, C. W. Gottschalk, W. E. Lassiter and M. Mylle: Micropuncture study of composition of proximal and distal tubular fluid in rat kidney. Amer. J. Physiol. **204**, 527 (1963).
71. Vurek, G. G., C. M. Bennett, R. L. Jamison, and J. L. Troy: An air-driven micropipette sharpener. J. appl. Physiol. **22**, 191 (1967).
72. —, and R. L. Bowman: Helium-glow photometer for picomole analysis of alkali metals. Science **149**, 448 (1965).
73. —, and S. E. Pegram: Fluorometric method for the determination of nanogram quantities of inulin. Anal. Biochem. **16**, 409 (1966).
74. Wahl, M., W. Nagel, H. Fischbach, and K. Thurau: On the application of the occlusion time method for measurements of lateral net fluxes in the proximal convolution of the rat kidney. Pflügers Arch. ges. Physiol. **298**, 141 (1967).
75. Walker, A. M., and J. Oliver: Methods for the collection of fluid from single glomeruli and tubules of the mammalian kidney. Amer. J. Physiol. **134**, 562 (1941).
76. Whittembury, G.: Site of potential difference measurements in single renal proximal tubules of necturus. Amer. J. Physiol. **204**, 401 (1963).
77. —, u. E. E. Windhager: Electrical potential difference measurements in perfused single proximal tubules of necturus kidney. J. gen. Physiol. **44**, 679 (1961).

78. WICK, T., u. E. FRÖMTER: Das Zellpotential des proximalen Konvoluts der Rattenniere in Abhängigkeit von der peritubulären Ionenkonzentration. Pflügers Arch. ges. Physiol. 294, R 17 (1967).

79. WIEDERHOLT, M., K. H. LANGER, W. THOENES und K. HIERHOLZER: Funktionelle und morphologische Untersuchungen am proximalen und distalen Konvolut der Rattenniere zur Methode der gespaltenen Ölsäule (split oil droplet method). Pflügers Arch. 302, 166 (1968).

80. WINDHAGER, E. E.: Peritubuläre Kontrolle der Natriumresorption im proximalen Tubulus. VI. Symposium der Gesellschaft für Nephrologie, Wien 1968 (in press).

81. —, E. L. BOULPAEP, and G. GIEBISCH: Electrophysiological studies on single nephrons. Proc. 3rd. int. Congr. Nephrol. Washington 1966, Vol. 1, pp. 35. Basel/New York: Karger 1967.

82. —, and G. GIEBISCH: Comparison of short-circuit current and net water movement in single perfused proximal tubules of rat kidneys. Nature (Lond.) 191, 1205 (1961).

83. WIRZ, H.: Druckmessungen in Kapillaren und Tubuli der Niere durch Mikropunktion. Helv. physiol. pharmacol. Acta 13, 42 (1955).

84. ZWIEBEL, R., B. HÖHMANN, P. FROHNERT und K. BAUMANN: Fluorometrisch-enzymatische Mikro- und Ultramikrobestimmung von Inulin und Glucose. Pflügers Arch. 307, 127 (1969).

Oscillatory Phenomena in a Porous, Fixed Charge Membrane

By **T. Teorell**

I. Introduction

The nerve action potentials were quite early recognized as oscillatory phenomena resembling certain rhythmical, chemical reactions at metal (Fe, Cr) interfaces (Ostwald, Lillie, Franck) and were described mathematically as 'relaxation oscillations' (Bethe, van der Pool, Bonhoeffer, Franck, Fitzhugh, and others). Many features exhibited by living, exitable tissues (nerve and heart) could be reproduced on these metallic systems, as all-or-non responses, threshold, bistability etc. The 'energy' driving the oscillations was obtained from electrochemical reactions.

In 1955 Teorell introduced a non-metallic system, built up by a porous membrane containing fixed charges (= 'ionic' membrane), which was surrounded by simple salt solutions like NaCl, KCl, etc. When an electrical potential gradient was applied damped or undamped oscillation phenomena occured (the 'energy was here supplied by a constant current from an external source). This model is called the 'membrane oscillator'. It can be 'stimulated' both by electrical current and by mechanical (hydrostatic) stimuli.

Further developments of the membrane oscillator has led to a proposal of an 'electrohydraulic excitability analog', which in a great many cases gives an excellent formalism for describing and reproducing actual biological excitability phenomena (threshold properties, refractoriness, make and break responses, abolition, 'frequency modulation', voltage clamp etc.).

The primary aim of the demonstration is to show a simple 'membrane oscillator' in actual operation (= *rhythmical* changes of *membrane potential*, transmembrane *resistance*, of *hydrostatic pressure difference* and *water flow*). The comments below will serve as an elementary introduction of the basic concepts on which the oscillatory mechanisms are supposed to rest. For the sake of compactness a *simplified mathematical description* will be given in order to show some general theoretical methods with which oscillatory phenomena can be handled ('non-linear mechanics', systems of linear or nonlinear differential equations).

II. Basic Concepts

It will be assumed that the student during the earlier lectures has become acquainted with the following basic concepts of membranology: *driving forces* [gradients of chemical potential, electrical potential and of

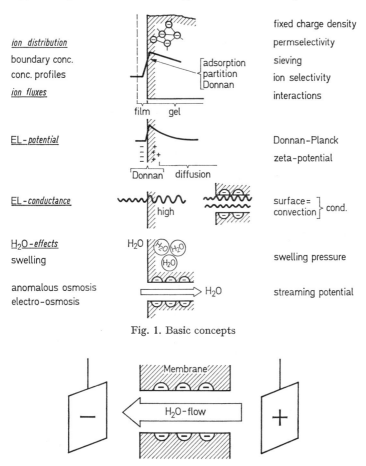

Fig. 1. Basic concepts

Fig. 2. Schematic diagram of electro-osmosis. An electrical potential across a membrane containing fixed ionic groups (here negative) creates a water flow moving towards the negative side

hydrostatic (osmotic) pressure differences], *membrane profiles* (i.e. spatial distribution of concentration, potential and pressure *within* the membrane), the general *nature of fixed charge (ion exchange) membranes, electrical membrane conductance* ('rectification') and *electrokinetical phenomena,* in particular *electro-osmosis* (c.f. Figs. 1 and 2).

9*

III. Principles of the 'Membrane Oscillator'

1. Consider a homogeneous, porous membrane separating two compartments (1) and (2), containing stirred solutions of different electric conductance, in the simplest case two solutions of a salt like NaCl, in the concentrations c_1, c_2 (see Fig. 3).

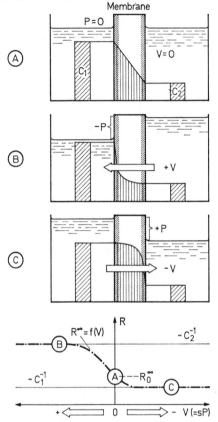

Fig. 3. Concentration profiles at different bulk flow velvcities (V)

The pores should be wide enough to permit filtration of the solutions, but narrow enough to support disturbance-free paths for diffusion. The pores must also contain 'fixed charges' on the pore walls (for instance arising from such fixed ions as carboxyl or amino groups in a protein membrane matrix). In the actual model it is a fritted glass disc (containing silicic acid).

The defined system is accordingly a simplified analog to a biological two-compartment system such as the 'inside' and 'outside' of a single cell membrane or tissue membrane.

2. An electrical current of a constant strength is assumed to pass across the membrane. In the case of a negatively charged membrane the high conductance compartment (1) is the negative pole and conversely the low conductance compartment (2) is the positive pole as shown in the following scheme:

Neg. pole	Solution (1) (high cond.)	negative membrane	Solution (2) (low cond.)	Pos. pole

For a positively charged membrane the direction of the steady current should be reversed.

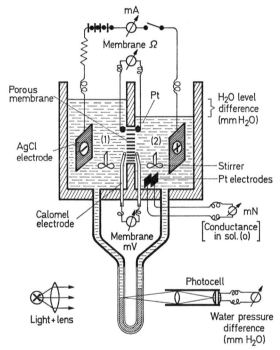

Fig. 4. Scheme of the experimental membrane oscillator

This set up corresponds to a 'membrane electrophoresis' arrangement, or rather, a system which would exhibit *electro*endosmosis. The actual source of the electric current across the membrane is immaterial for the theory, in practical experiments, the current has been introduced from an external D.C. source via AgCl electrodes directly or indirectly placed in contact with solutions (1) and (2) as in the scheme (Fig. 4).

An electroosmotic flow of the solution will now take place, tending to increase the fluid volume in compartment (1) and to decrease the volume of compartment (2). If both, or at least one of the compartments, are of a small, finite volume and *open* in the sense that free surface levels can be established, it is obvious that the electro-osmotic flow will induce a *hydrostatic pressure difference, P*, across the membrane.

One now applies the well-known expression for *electro-osmotic fluid transport*, which is the sum of a potential term and a pressure term:

$$V = l \cdot E - s \cdot P . \tag{1a}$$

This is called 'Wiedemann's law'. Here E is the electric potential difference across the membrane, l signifies an '*electro-osmotic permeability coefficient*' directly related to the fixed charge density (ωX) of the membrane and s, the '*hydrostatic permeability coefficient*' for the liquid flow $(l = - sF\omega X)$. F is the Faraday constant.

3. From the simple geometrical lay-out of the model system it follows that at any instant the mean linear *velocity of flow, V*, of the fluid across the membrane will be related to the *hydrostatic pressure difference, P*, across the membrane as

$$V = q(dP/dt) \tag{1b}$$

where q is the 'geometry' coefficient and t is the time.

4. The fundamental concept in the oscillatory mechanism centers around the *changes of the electrical membrane resistance, R*, as related to the velocity of fluid movement, V. This resistance is in turn dependent on the course of the 'concentration profile' across the membrane. When $V = 0$ this profile will conform to the requirements of diffusion, which, in the final *steady state*, is a *straight* line. If $V \neq 0$ the profile will be distorted and form a *convex* or *concave* curve dependent on the direction and magnitude of V as depicted in Fig. 3. In the extreme cases of very high V the membrane will be filled up with either the solution (1) or (2) and hence the resistance will be inversely proportional to the concentration of either of the surrounding solutions, depending on the flow direction. Thus it is possible to predict without calculations, that the *steady state* membrane *resistance* R^∞ will be a function of V, varying around an intermediate resting value of R_0^∞ at $V = 0$ with a low and a high limiting value at extreme flow velocities. Hence, it can be understood that $R^\infty = f(V)$ is an *S-shaped function*, with an approximate straight line segment over a region of small values of $\pm V$ near the resting value R_0^∞ (see lower section in Fig. 3). In this region the function $R^\infty = f(V)$ can be simplified to the *linear* relation

$$R^\infty = rV + R_0^\infty \tag{2}$$

where r, the slope, or inclination, is a constant coefficient for a given system [the exact equation can be found in Teorell, 1959 (II)].

5. A change of V cannot induce an instantaneous attainment of the corresponding steady state value of R^∞. On the contrary, the *adjustment*

from one concentration profile to another requires time due the relative slowness of the diffusion process, which is superimposed on the bulk flow. The time delay expressed as the *rate of change dR/dt* of the instantaneous membrane resistance R occuring at any given V, is a complicated function. It seems plausible, however, that one can again resort to a linear approximation for small V's, i.e. R values not too far off from the resting value of R_0^∞. Therefore, for a given V, one can assume that

$$dR/dt = -k(R - R^\infty) \tag{3}$$

i.e. *the rate of change of the membrane resistance is directly proportional to the 'deviation from the steady state'*. The coefficient k is regarded as constant for a given electrolyte-membrane system (but is dependent on the same membrane factors as r, i.e. thickness, diffusion constant etc.).

6. The relation between the electric current density i, the transmembrane potential E and the hydrostatic pressure P has for long been known to be $i = E/R + l \cdot P$ (which can be derived by aid of irreversible thermodynamics). In relatively wide pores and with not too small i the contribution of the 'streaming current', $l \cdot P$, can be neglected and the validity of *Ohm's law* assumed i.e.

$$E = i \cdot R . \tag{4}$$

IV. Mathematical Equations

1. *List of notations.*

c = concentration
E transmembrane potential (instantaneous, approximately $E = i \cdot R$)
E^∞ = transmembrane potential (steady state)
E_0^∞ = transmembrane potential at $V = 0$ (steady state)
R = resistance (instantaneous)
R^∞ = resistance (steady state)
R_0^∞ = resistance at $V = 0$ (steady state)
P = hydrostatic pressure difference
V = (linear) velocity of 'water' flow
i = current density
s = hydraulic permeability constant
l = electro-osmotic permeability constant $[= -F \cdot s\,(\omega \bar{X})]$
(ωX) = membrane fixed charge density ($\omega = \pm 1$, 'charge sign')
k = 'delay constant' (reciprocal time constant of R change)
q = 'geometry' constant
r = slope of constant slope segment of $R^\infty = f(V)$
t = time
x = $(E - E_0)$
y = V

The 'parent constants' i, s, l, k, r, and q can be combined to the following 'lumped' constants:

$$\alpha = (kir) \qquad \beta = (-k) \qquad \gamma = (klir - s/q) \qquad \delta = (-kl) .$$

2. Summarizing the *basic equations* from the previous paragraphs, III, 3 — 6, one obtains the following systems ($\dot{P}$, $\dot{R}$, $\dot{E}$, and $\dot{V}$ signify the derivatives with respect to time dP/dt etc.):

$$V = l\cdot E - s\cdot P \tag{1a}$$

$$V = q\cdot \dot{P} \tag{1b}$$

$$R^\infty = r\cdot V + R_0^\infty \tag{2}$$

$$\dot{R} = -k(R - R^\infty) \tag{3}$$

$$E = iR . \tag{4}$$

An *operational scheme* of the 'couplings' in the equation system is given in Fig. 5.

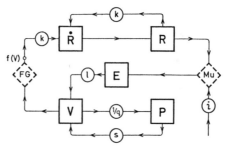

Fig. 5. Operational Scheme of Eqs. 1—4

3. *Solution*: We will now elect to solve the system of Eqs. (1 — 4) for E and V with respect to time (t).

Combining (2) and (3) leads to

$$\dot{R} = -k(R - r\cdot V - R_0^\infty) .$$

Multiplying by i and making use of (4) yields

$$\dot{E} = (kir) V + (-k)(E - E_0^\infty) . \tag{5}$$

To obtain an expression for $\dot{V}$ one differentiates (1a) with respect to time

$$\dot{V} = l\cdot\dot{E} - s\cdot\dot{P} \tag{6}$$

and substituting in (6) the value of $\dot{E}$ from (5) and $\dot{P}$ from (1b) respectively:

$$\dot{V} = (klir - s/q) V + (-kl)(E - E_0^\infty) \tag{7}$$

introducing finally the new variables $(E - E_0^\infty) = y$ and $V = x$ one obtains from (5) and (7)

$$dy/dt = \alpha x + \beta y \tag{8}$$
$$dx/dt = \gamma x + \delta y$$

where $\alpha = (kir)$; $\beta = (-k)$; $\gamma = (klir - s/q)$ and $\delta = (-kl)$.

The solution of the differential equation system of the type (8) gives under certain conditions integral curves showing an *oscillatory time-dependent variation of x and y, i.e. of the velocity of the fluid V and the 'driving' electrical potential E*. Damped, undamped or 'growing' sinuoidal oscillations of E and V can result when $[(\beta - \gamma)^2 + 4\alpha\delta)] < 0$, their character dependent on whether the 'damping-factor' $(\beta + \gamma)$ is < 0, equal to 0 or > 0 respectively (see such text-books as LANGER, 'Ordinary differential Equations', New York, 1954 or McLACHLAN 'Ordinary non-linear Differential Equations', Oxford, 1956).

It is easy to show that a corresponding rhythmicity applies to the pressure difference P and the transmembrane resistance R.

Note on more general treatments: The simplified case analyzed above was linear. *Non*-linear cases can be solved graphically by the 'isocline method' outlined by TEORELL, 1959 II, or by analog machine technique. — An instructive, somewhat different approach is presented by FRANCK, 1963.

V. Demonstration of the 'Membrane Oscillator'

1. *The construction of the apparatus* was shown in Fig. 4. It consists of two Perspex chambers between which a porous glass or a porcelain membrane is clamped. The membrane discs are commercial types used for filtration, about 1.5 mm thick, with an area of $1 - 4$ cm² and with pore diameter of about 1 μ. One chamber is filled with 100 mM NaCl, the other with 10 mM NaCl. The volumes are $10 - 20$ ml.

A constant *D. C.* current is supplied from a constant current source through a large AgCl electrodes, either inserted directly in the chambers, as indicated in the figure, or placed in special electrode vessels. The latter arrangement is preferable for long lasting experiments.

The *current strength*, of the order of $5 - 40$ mA, is recorded on a mA-meter. The *membrane potential*, in the main an ohmic voltage drop $(E = i \cdot R)$, is recorded by built in calomel electrodes, with tips close to the membrane surfaces, connected to a high impedance voltmeter. Similarly, the *membrane resistance* is measured between two platinum electrodes by means of a conductance meter. The *hydrostatic pressure difference* between the two chambers, which arises during the experiment, is recorded directly as differences between the free liquid levels or by means of a differential manometer (preferably a differential pressure transducer connected to a meter).

2. During *the demonstration* continuous readings will be taken of the *transmembrane potential*, the *transmembrane resistance*, and the *hydrostatic pressure difference*. Flux measurements of ions will be omitted.

Two types of experiments will be run, one with a low, one with a higher current density, yielding a *damped* and an *undamped*, sustained

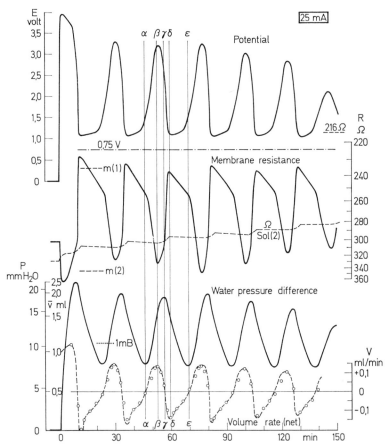

Fig. 6. Experimental results of a system N/10 NaCl/porous porcelain membrane/ N/100 NaCl at 25 ma

oscillation, respectively, (possibly the effect of a 'pressure stimulus' will be tried by plunging a fixed glass rod into one of the solutions, thus suddenly raising its free surface level. Other possible experiments are 'pressure clamping' and 'voltage clamping', which will be described).

Membrane interactions can be demonstrated by addition of methylene blue. This dye (basic) adsorbes to the negative, fixed silicic acid residues and diminishes the membrane charge, thus 'poisoning' and 'killing' the oscillations. Increase of alkalinity, on the other hand, enhances the oscillations.

3. A *typical result* is depicted in Fig. 6. In this also the flow velocity, V, is recorded. This can be obtained as a derivative of the time curve of the pressure difference, P.

VI. Summary

A model and a tentative theory has been presented showing that an electrolyte membrane system of a biologically possible type can exert oscillatory fluid transport across the membrane if this is subject to the passage of a constant current. The oscillations can be damped or continuously undamped. The main mechanism of this process arises from cyclical changes in the electrical resistance of the membrane induced by a fluid flow arising from electro-osmosis (owing to the presence of 'fixed charges' in the membrane). Concomitant phenomena are oscillations of the membrane potential and of a hydrostatic pressure difference across the membrane.

References

A. Papers describing the membrane oscillator, experiments, and theories

TEORELL, T.: Acta physiol. scand. 31, suppl. 268 (preliminary communication) (1954).
— A contribution to the knowledge of the rhythmical transport processes of salt and water. Exp. Cell Res. Suppl. 3, 339 (1955).
— Transport processes in membranes in relation to the nerve mechanism. Exp. Cell Res. Suppl. 5, 83 (1958).
— On oscillatory transport of fluid across membranes. Acta Soc. Med. upsalien. 62, 60 (1957).
— Rectification in a plant cell (Nitella) relation to electro-endosmosis. (Compares effects of square wave current on Nitella cells with corresponding results on the membrane oscillator.) Z. physik. Chem. NF 15, 25 (1958).
— Electrokinetic membrane processes in relation to properties of excitable tissues. I. Experiments on oscillatory transport phenomena in artificial membranes. (A full technical description of the membrane oscillator; damped and undamped oscillations; current and pressure stimuli.) J. gen. Physiol. 42, 831 (1959).
— Electrokinetic membrane processes in relation to properties of excitable tissues. II. Some theoretical considerations. (Basic equations, description of graphical methods of solving non-linear differential equations, i.e. the 'isocline method'.) J. gen. Physiol. 42, 847 (1959).
— Oscillatory electrophoresis in ion exchange membranes. (Membrane of granulated ion exchanger, oscillations with identical solutions inside and outside.) Arkiv för kemi (Roy. Swed. Acad. Science) 18, 401 (1961).
— The ion flux across membranes during electro-diffusion and convection. (Theory for ion flux in single salt/porous membrane systems.) Acta physiol. scand. 62, 293 (1964).

B. The electrohydraulic excitability analog

TEORELL, T.: Biophysical aspects on mechanical stimulation of excitable tissues. (Analog computation on a modified membrane oscillator, demonstrating features of pressure stimuli.) Acta Soc. Med. upsalien. 64, 341 (1959).

— Application of a voltage-clamp to the electro-hydraulic nerve analog. (Analog computation of voltage clamp responses, discussion of nerve voltage clamps.) Acta Soc. Med. upsalien. **65**, 231 (1960).
— An analysis of a current-voltage relationship in excitable Nitella cells. (Nitella can give 'dynamic' V-I-characteristics, which are analyzed in terms of the electrohydraulic theories.) Acta physiol. scand. **53**, 1 (1961).
— Some biophysical considerations of presso-receptors. (Analysis of the effects of dynamic pressure stimuli, mathematical and graphical methods, analog computation, discussion of presso-receptor properties.) Arch. int. Pharmacodyn. **140**, 562 (1962).
— Electrokinetical considerations of mechano-electrical transduction. (Further developments of the electrohydraulic pressoreceptor analog with a pronounced 'frequency modulation', discussion of a relation impulse frequency *vs* stimulation intensity, relations between generator potentials and action potentials.) Ann. N. Y. Acad. Sci. **137**, 950 (1966).

C. General reviews of the membrane oscillator and excitability

Teorell, T.: Excitability phenomena in artificial membranes. (An introduction and review.) Biophys. J. **2**, No. 2, part 2, suppl., 27 (1962).

D. Some references to recent works by other authors

Aranow, R. H.: Periodic behavior in charge membranes with physical and biological implications. Proc. nat. Acad. Sci. (Wash.) **50**, 1066 (1963).
Franck, U.: Über das elektrochemische Verhalten von porösen Ionenaustauschermembranen. Ber. dtsch. Bunsenges. physik. Chemie (Z. Elektrochem.) **67**, 657 (1963).
Kobatake, Y., and H. Fujita: Flow through charged membranes. I. Flipflop currents vs. voltage relation. J. chem. Phys. **40**, 2212 (1964).
— — Flows through membranes. II. Oscillation phenomena. J. chem. Phys. **40**, 2219 (1964).

E. Some general references to fixed charge (ion exchange) membranes

Helfferich, F.: Ionenaustauscher, Bd. I. (A handbook with complete literature references, English translations available.) Weinheim (Bergstraße): Verlag Chemie GmbH 1959.
Schlögl, R.: Stofftransport durch Membranen. Darmstadt: Dr. R. Steinkopff Verlag 1964.
Teorell, T.: Zur quantitativen Behandlung der Membranpermeabilität. (Basic mathematical theory of fixed charge membranes.) Z. Elektrochem. **55**, 460 (1951).
— Transport processes and electrical phenomena in ionic membranes. (Monograph covering membrane theories, transport kinetics, distribution equilibria, and electrical membrane phenomena.) Progr. Biophys. **3**, 305 (1953).
— Transport phenomena in membranes. (A general lecture on membrane phenomena.) Farad. Soc. Disc. No. **21**, 1 (1956).

Bimolecular Lipid Membranes: Techniques of Formation, Study of Electrical Properties, and Induction of Ionic Gating Phenomena

By **P. Mueller** and **D. O. Rudin**

I. Introduction

a) Properties of Lipid Bilayers. Bimolecular membranes formed from cellular lipids have many properties of cell membranes [1 – 3]. They have the same values for thickness (75 ± 15 Å) [1, 3, 4], electrical capacitance (0.4 – 1.2 μF depending on lipid type) [1, 5], dielectric strength (5×10^5 V cm^{-1}) [1 – 3], water permeability (1 μ min^{-1} atm^{-1} by osmometry) [3] and surface tension (between 0 and 5 dynes cm^{-1}) [3, 6].

One outstanding difference is their high electrical resistance which at 10^8 Ω cm^2 is $10^5 - 10^7$ times higher than that of most cell membranes. However, this high resistance can be lowered into the physiological range by a few classes of compounds which adsorb to the membrane from the water phase and increase the ionic permeability by many orders of magnitude. These adsorbates include certain detergents [3], some amphophilic chelators, selected macrocyclic antibiotics, porphyrins and a proteinaceous compound of cellular origin called EIM (excitability inducing material) [1, 7].

The last compound is of special interest because it forms cation conducting membrane channels which in the presence of ionic gradients generate resting potentials of up to 160 mV and have the electrokinetic properties attributed to the potassium channels in nerve. Moreover, in membranes made from beef heart lipid a portion of these channels can couple with protamine and the system then generates action potentials with electrokinetics and pharmacological responses indistinguishable from those of many cells.

The action potentials are primarily the result of the voltage controlled permeability changes, or gating of the EIM channels, which in the absence of a resting potential generate two negative resistance regions in the steady state current-voltage curve and the corresponding time-voltage transitions under constant currents [7] (Figs. 3, 4). The gating properties are qualitatively independent of the lipid composition and are demonstrated here in membranes made from beef brain lipids, which are easier to make and are stronger than those made from beef heart lipid.

b) Principles of Membrane Formation. Bimolecular lipid membranes are formed in water by a process analogous to the formation of 'secondary black' bimolecular soap films in air [1]. A soap film held on a loop showing interference colors will spontaneously thin to the limiting stable unit for the smectic phase in which two monolayers are joined at their charged groups while the hydrocarbon chains project perpendicularly to the plane of the film into the air. These films are between 40 and 60 Å thick, reflect very little light and have been called 'secondary black'. Lighter gray areas called 'primary black' result if thinning is not complete. They are unit multiples of secondary black films [8].

In the lipid membranes the charged groups face the water while the two planes of the hydrocarbon chains are joined by van der Waals forces. Such membranes can be formed from a variety of single or mixed lipids provided appropriate solvents and nonpolar additives are used to increase the liquidity and lower the surface tension of the bulk lipid phase while the bilayer is forming. The lipid solution is spread as a film on a small (1 mm diam.) aperture in teflon, polyethylene or other inert plastic by means of a soft brush. The formation of secondary black often starts at several spots and the black regions look initially like small holes which grow, merge, and finally occupy the entire membrane area. The bulk lipid is squeezed toward the rim or into droplets which initially remain suspended in the black area but gradually move to the side.

When such membranes are made so that they separate two aqueous compartments, their electrical properties can be studied as a function of their own composition and that of the aqueous phase.

II. Materials and Instruments

a) Instruments.

1. Membrane chamber and accessories as in Fig. 1. Should be mounted on vibration free surface.

2. Stimulator and cathode follower as in Fig. 2. Battery operated. Should be connected to power at all times. Keep 6 V battery charged and check two 45 V batteries at regulator intervals.

3. Recording instruments such as oscilloscope and/or strip chart recorder.

4. Heat lamp;

5. Two calomel electrodes (Beckman);

6. KCl-Agar bridges made from 1 mm diam. polyethylene tubing, bent to connect the two holes (f) with dish and cup. The bridges are filled with solution No. 8 and stored in covered Petri dish.

7. Sable hair brush No. 0;

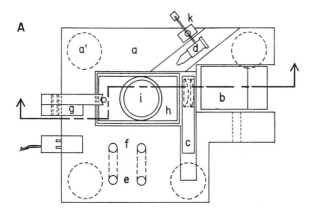

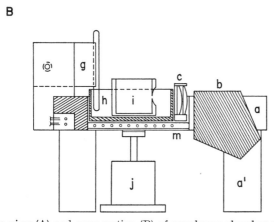

Fig. 1. Top view (A) and cross section (B) of membrane chamber and accessories. (a) Body (black plexiglas) supported on 4 feet (a'). (b) Pentaprism. (c) Lens system. Two lenses (1) Zeiss achromat — diameter = 16 mm, F. L. = 36mm — and (2) plano convex lens — diameter = 16 mm, F. L. = 40 mm — mounted in a sliding bar of black plexiglas. (d) Light bulb (G. E. TL 222) and socket mounted on k. (e) Holes to receive calomel electrodes. (f) Holes for KCl bridges. KCl-agar filled polythylene tubing connects dish and cup with the holes; (e) and (f) are connected and filled with saturated KCl. (g) Thermometer and holder. (h) Membrane dish (plexiglas). (i) Membrane cup of polyethylene or teflon, either injection molded or machined. (j) Stirring motor (Solar drive motor, International Rectifier Corp.) and magnet, mounted on lead disc. (k) Cortical electrode holder (Grass Instrument Co., Quincy, Mass.) inserted into (a). (m) Heating plate. Controls for light, heat and stirrer, two 50 Ω potentiometers and switch — input 6.0 V — are mounted separately

8. Stirring bars, 8 mm length of 0.5 mm diam. monel wire. Store in contact with small magnets.

9. Filter paper.

b) Solution and Chemicals.

1. 25 cc 4% mixed lipids in 2:1 chloroform:methanol extracted from beef brain according to the following procedure:

a) Mix 50 g fresh beef white matter and 1000 ml $CHCl_3/CH_3OH$ 2/1 V/V (spectral grade) in a high speed blender for 2 min.

b) Remove residue by centrifugation.

c) Add 100 ml of water and mix.

d) Take to dryness by vacuum pump (24 h).

e) Dissolve the residue in 100 ml of chloroform and filter.

f) Add $\frac{1}{2}$ volume of methanol.

g) To 100 ml of f) add 20 ml 0.1 N NaCl, shake, centrifuge, discard upper phase, filter lower phase three times and add methanol equal to $\frac{1}{3}$ of the remaining volume. Store dark at room temperature. Solution is stable for months. Lipids extracted this way during winter and spring give more stable membranes than those from summer and fall.

2. α-Tocopherol, 10 g.

3. Cholesterol, 2 g.

4. $CHCl_3:CH_3OH$, 2:1, 50 ml.

5. 0.1 M NaCl, 1000 ml.

6. 0.1 M Histidine chloride, pH 7, 50 ml.

7. Saturated KCl, 100 ml.

8. 3% Agar in 3 M KCl, 250 ml.

9. 5.0 g crude EIM obtained from *enterobacter cloacae* ATCC 961 by the following method:

a) Inoculate 50 ml thioglycolate broth with 5 mg lyophilized *enterobacter cloacae* ATCC 961. Grow overnight at 35 °C.

b) Stir a) into 1000 ml of egg white, previously titrated to pH 7 with HCl, and incubate overnight at 35 °C.

c) Dry on flat pan with warm air and store dry at room temperature.

The preparation is stable in this form for 1 − 2 years. EIM can also be consistently obtained by growing *enterobacter cloacae* ATCC 961 in artificial media, or by extracting the washed bacteria at high pH or by extracting yeast acetone powders with water.

10. To 500 ml of solution No. 5 add 25 ml of solution No. 6, boil to degas, cool and fill into bottle or equivalent. Prepare fresh every day.

11. Lipid membrane solution: To 1 cc of solution No. 1 add 400 mg of compound No. 2 and 30 mg of compound No. 3; mix well and keep closed when not in use. The solution deteriorates within days to 1 week.

If membranes do not form readily or tend to break, prepare fresh as directed.

12. 5% aqueous solution of No. 9, pH 7, 1 ml. Store in refrigerator and prepare every other day.

c) Glassware.

1. Volumetrics: 1 cc, 4; 10 cc, 1; 50 cc, 3; 100 cc, 1; 250 cc, 1; 500 cc, 1; 1000 cc, 1.
2. Erlenmeyer flask, 50 ml, 1.
3. Beaker, 50 ml, 1.
4. Disposable pipettes.

III. Execution

a) Membrane Formation.

1. Precondition a clean membrane cup* by wetting the area around hole from both sides with solution No. 1 by means of hair brush. Insert membrane cup into plexiglas dish and then insert dish into holder (Fig. 1).
2. Fill dish and cup with solution No. 10, 3 mm below rim.
3. Insert the thermometer.
4. Turn on light bulb or heater in membrane chamber.
5. Adjust position of cup so that hole points toward light and is in focus when viewed through prism from above.
6. Put stirring rod into cup and turn on stirrer; make certain that stirrer runs smoothly without vibrating the cup.
7. Fill the two holes in the holder next to the dish with saturated KCl. Insert calomel electrodes into outer holes and KCl bridges from center holes to dish and cup.
8. Electrical checks:

a) Check the polarity of the recording and stimulating equipment by connecting a small battery to the electrodes and observing the recorder deflection.

b) With input resistance $(S - 3$, Fig. 2) at 100 MΩ, check the electrode potential by switching $S - 5$ between open and operate. This connects and disconnects the electrodes at the input. There should be not more than 5 mV potential difference in the two switch positions.

c) Check electrode resistance. Set $S - 5$ to operate, R_i to 0.1 MΩ, V_i to 10. Press $S - 1$. The recorded potential should be < 25 mV.

---

* Washing procedure for membrane cups:
a) Soak several hours in 5 N NaOH.
b) Scrub region around hole in and out with cotton swab and NaOH.
c) Rinse 10 × in distilled water.
d) Soak in 1 N HCl for 1 – 2 h.
e) Rinse 10 × in distilled water.
f) Soak in deionized water several hours.
g) Let dry.

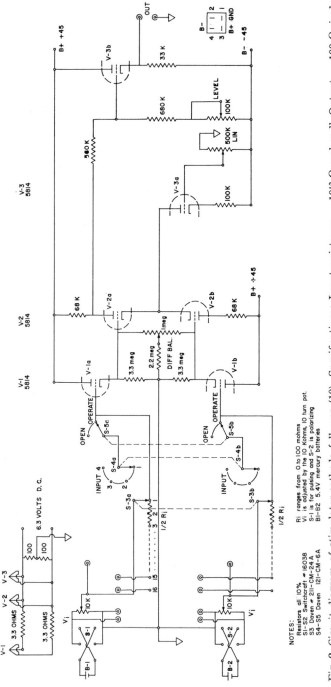

Fig. 2. Circuit diagram of stimulator-cathode follower (10). Specifications: Input resistance = 10^{12} Ω push pull. Output = 100 Ω single ended. Drift = 3 mV/hr. Noise = 50 μV rms. Inphase rejection = 1000/1. High frequency response = 50% at 10 KC. Gain = ~ 2. Stimulator: Two manually operated, independent circuits with maximum voltage of 5.4 V each. Series resistors variable from 0.1 to 1000 MΩ in 15 steps. The circuit was designed by Y. C. Chock and is manufactured by Electronics for Life Sciences, Rockville, Maryland, U.S.A.

d) Check cathode follower balance. Set R_i to 1000 MΩ (10^9 Ω). Set V_i to 1000. Press $S - 1$. Potential should read < 10 mV. Correct by adjusting differential balance or replacing tube 1.

e) With R_i at 10^9 Ω, calibrate your recording circuit by switching $S - 5$, Fig. 2 open and applying 54 mV to the cathode follower input by setting the 10 turn potentiometer (V_i) to position 10 and pressing $S - 1$ briefly.

9. Apply heat lamp until the temperature reaches 35 °C. Keep this temperature while membrane forms.

10. Dip tip of hair brush into lipid solution. Do not fill brush completely. Remove excess by touching clean filter paper.

11. Under visual control, apply lipid film to the hole in the cup by wiping the brush gently from one side of the hole to the other. Do not wipe back and forth excessively.

12. Important: Make certain that you see the interference colors of the light reflected from the film. If not, adjust the position of cup and light with a film in position until bright interference colors appear. The cup can be rotated and the light can be moved up and down.

13. If the film breaks, clean the brush by dipping into 50 cc of chloroform-methanol (solution No. 4). Wipe dry on clean filter paper and repeat steps No. 11 and 12.

14. If film breaks more than five times before EIM has been added, wipe area around cup hole several times with brush filled with chloroform-methanol. Clean the brush each time as in step No. 13 and finally remove excess of chloroform-methanol on hole by wiping with dry brush. If membrane still keeps breaking or if it breaks after EIM was added to the cup, remove dish, cup and KCl bridges, clean the thermometer and proceed from step No. 1.

15. If the film does not break and was properly positioned, the formation of secondary black can now be observed as described above. Notice the change of interference colors and the speed with which the black area increases. Membranes that turn black too fast, i.e. within a few seconds, tend to be unstable. In this lipid the rate of black formation can be decreased by lowering the temperature or increasing the lipid concentration in solution No. 1. If no black appears within 5 − 10 min, the process can be initiated by carefully touching the colored film with one hair of the brush. This often creates a focus from which the phase transition starts.

b) Measurement of Membrane Resistance and Capacitance.

Principles: Fig. 3 shows the essential elements of the measuring and recording circuit. A selected voltage, V_i is applied from a calibrated source to the input resistor, R_i, and the membrane resistor, R_M. The portion of this voltage appearing across the membrane V_M, is recorded with a cathode follower between the two calomel electrodes and is given by

$$V_M = \frac{V_i\,R_M}{R_i + R_m}.\tag{1}$$

Since V_i and R_i are known and V_M is measured, R_M can be calculated from

$$R_M = \frac{V_M\,R_i}{V_i - V_M}.\tag{2}$$

In practice, the measurement is most easily made if R_i is adjusted so that $1.3 < V_i/V_M < 10$. Typical values of the specific membrane resistance lie between 10^7 and $10^8\ \Omega\ cm^2$, and since the actual membrane area is approximately 0.75 mm^2, the measured values of R_M are in the order of $10^9\ \Omega$.

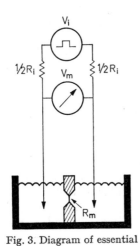

Fig. 3. Diagram of essential stimulating and recording circuit elements. V_i = stimulating voltage, R_i = series input resistor, V_M = recorded membrane potential, R_M = membrane resistance

Procedure:

1. Set $S - 5$ to 'operate'. Set R_i to $10^9\ \Omega$ and V_i to 10. Press $S - 1$ until the recorded potential is steady. This potential is V_M.

2. If V_M is < 5 mV, decrease R_i until V_M reads anywhere between 5 and 40 mV.

3. Calculate R_M from Eq. (2) inserting the values of V_i, V_M, and R_i. R_M often keeps increasing for several minutes after the membrane turns black. This is probably due to a continued diffusion of solvent (chloroform-methanol) into the water. If R_M stays $< 10^8$, check for faulty cup insulation or prepare new lipid solution.

c) Measurement of Membrane Capacitance.

Principles: An applied potential charges the membrane capacitance. When the external voltage is instantaneously removed, the capacitance discharges and the membrane potential varies in time according to

$$V_M = V_{M_0}e - \frac{t}{RC_M}.\tag{3}$$

$V_{M_0}e$ is the initial voltage, i.e. the membrane potential at the time when the external current was interrupted. R is the resistance through which the capacitance, C_M, discharges and t is the time. C_M can, therefore, be calculated from:

$$C_M = \frac{t}{R} \frac{1}{\ln V_{M_0}/V_M}.$$ (4)

If t is measured when $V_M/V_{M_0} = 1/e = 0.36$ and called τ, Eq. (4) reduces to

$$C_M = \frac{\tau}{R}.$$ (5)

$\tau = RC_M$ is called the time constant. C_M is in parallel with R_M and R_i and can be discharged through R_M and R_i. Therefore

$$\frac{1}{R} = \frac{R_i + R_M}{R_M R_i}$$ (6)

and

$$C_M = \tau \frac{R_i + R_M}{R_M R_i}.$$ (7)

Procedure:

1. Apply 50 mV across the membrane by pressing $S - 1$ and adjusting V_i. Release $S - 1$ and measure τ at the point where the voltage has dropped to 19.5 mV.

2. Insert τ into Eq. (7) and calculate C_M. Convert to μF cm^{-2} after estimating the black area. This value is the low frequency capacitance. Sine wave analysis has shown that the capacitance of the unmodified membrane is independent of frequency between 10^{-5} and 10^6 and voltage up to 50 mV [5].

d) Measurement of Dielectric Strength.

Procedure:

1. Reduce the gain of the recorder. Increase the applied membrane potential until the membrane breaks.

2. Convert to V/cm assuming a membrane thickness of 70 Å. The dielectric strength is a function of the duration of the applied voltage as well as lipid composition and composition of the aqueous phase.

e) Lowering of Membrane Resistance by EIM.

Procedure:

1. Make a new membrane and measure the resistance. It should be $> 2 \times 10^8 \, \Omega$.

2. Raise the temperature to 40 °C and keep it constant during the rest of the experiment.

3. Check stirrer for proper action.

4. Add 0.05 ml crude EIM (solution No. 12) (1 drop) to the cup. Touch a connection to ground while adding or withdrawing solution. Withdraw equal amount and check to make sure membrane is flat.

5. Apply 5 sec pulses every 30 sec and while depressing $S - 1$ adjust V_i and R_i so that membrane potential reaches 25 mV during each pulse. Avoid settings of V_i above 500 because they tend to cause artifacts (see section II A 8 d).

6. Note values of V_i and R_i together with time after addition of EIM. Continue to take readings for 5 min.

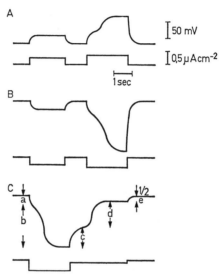

Fig. 4. Membrane voltage (upper trace) in response to applied constant currents (lower trace). A — Outward currents, B and C inward currents (inside = side of EIM). The potential is referred to the outside which is taken as zero. Note the thresholds and the asymmetry of the two polarities. In C the reduction of the current caused an inverse transition. The letters in C indicate voltage steps which correspond to the voltage steps in the I(V) curve in Fig. 5

7. If R_M does not decrease, check temperature, stirring action and if membrane is black and flat. Add two more drops of EIM if necessary and check R_M for another 5 min.

8. If R_M decreases below $10^6 \, \Omega$ within 5 min and the membrane breaks during this time, proceed from step No. 1 but use a $10 \times$ diluted EIM solution.

9. Plot membrane conductance $G_M = 1/R$ as a function of time and calculate dG_M/dt cm^{-2} for this EIM concentration in μmhos min^{-1} cm^{-2} μg^{-1} ml at 3 min after application of EIM. The volume of the cup is 4 ml.

For a given lipid batch and membrane solution additives this value is an approximate measure of the specific activity of the EIM sample and can be used for the assay of EIM during purification. Compare the value obtained with that of highly purified EIM which is ca. 10^3 μmhos min^{-1} cm^{-2} μg^{-1} ml (9).

f) Time-Voltage Characteristic of Membrane with EIM.

Procedure:

1. Proceed as in section E (steps No. 1 — 6).
2. After the membrane resistance has dropped below 50 MΩ, adjust R_i so that V_i at a setting of 50 gives about 10 mV membrane potential.
3. Apply inward current square pulses of 5 sec duration at 10 sec intervals. 'Inside' is the compartment containing EIM. Increase V_i from pulse to pulse until a nonlinear voltage increase appears during the pulse (see Fig. 4). Interrupt the pulse if the membrane potential increases fast above 100 mV and reduce R_i until the steady state potential after the nonlinear voltage increase is < 100 mV. This shifts the operating conditions more towards 'constant voltage' and reduces membrane breakage.
4. Find the threshold for the voltage transition. Notice the latency near the threshold and observe the effect of varying the pulse interval on the latency.
5. Check the membrane resistance during and after the transition by applying short pulses with $S - 2$.
6. Use $S - 2$ to initiate a transition with inward current. Keep current on and apply increasing outward currents (3 sec) from $S - 1$ until a reverse transition occurs (Fig. 4c).
7. Repeat step No. 3 with outward currents. Notice the different threshold and transition speed.
8. If membrane breaks, try again.

g) Current-Voltage Characteristics.

Procedure:

1. Proceed as in section F (steps No. 1 — 4, 6, 7). Use both polarities.
2. Record V_i and R_i for each steady state membrane potential. Obtain values between ± 120 mV.
3. Calculate membrane current I_M from

$$I_M = \frac{V_i - V_M}{R_i} \tag{8}$$

and plot I_M cm^{-2} against V_M as in Fig. 5. This is the $I(V)$ curve under approximate constant current conditions and is difficult to obtain because the membranes tend to break at the high voltages. The curve shows an upward curvature at potentials above 60 mV which indicates a secondary resistance decrease. This feature of the $I(V)$ curve is the cause of anomalous rectification which occurs if a sufficiently large resting potential drives the membrane resistance beyond its maximum.

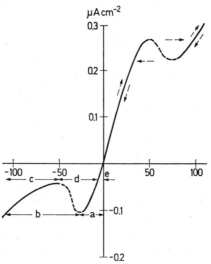

Fig. 5. Current voltage curve of a membrane with adsorbed EIM and in equal ion concentrations on both sides. Inside is defined as the side to which EIM was added and inward currents are plotted downwards. The voltage steps indicated by the letters correspond to the voltage steps that occur in the constant current-time-voltage curve (Fig. 4). Under constant current conditions the membrane voltage varies along a path indicated by the arrows and the region of negative slope cannot be measured

IV. Theory

a) Calculation of the I(V) Curve.

The $I(V)$ curve can be calculated in first approximation under the following assumptions:

a) EIM forms a number of membrane channels, n (see Fig. 6).

b) Each channel has two gates, R_1 and R_2, in series.

c) Neglecting state c in Fig. 6, the gates can have two configurational states with a low or high resistance, a or b, i.e. open or closed.

d) The state of each gate depends on the membrane potential, the free energy difference, ΔF, between the two configurations and the thermal energy. Normally, the free energy difference between the two states is such that most of the gates are in their low resistance state at zero potential.

e) The transition between the two configurations is a first order reaction.

f) The two gates in series in each channel are driven into their high state by voltages of opposite sign.

Considering only gates R_1, if p is the fraction of gates in the high resistance state, b, and $1 - p$ the fraction in state a, then the rate of change of p,

$$dp/dt = k_1(1 - p) - k_2 p .\qquad(9)$$

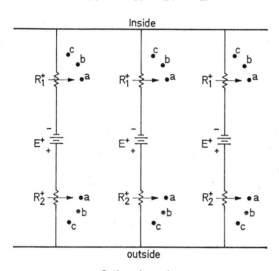

Inside

outside

Cation channels
(EIM–Lipid)
Channel Density=n/cm^{-2}

Fig. 6. Equivalent circuit of three membrane channels formed by EIM. There are n cm^{-2} of such channels. The gates R_1 and R_2 act as internal resistance of the emf, E^+. Each gate can exist in three resistance states, the state of each gate being determined by membrane potentials of opposite sign. State a is one of low resistance at zero potential, state b has a high resistance at ± 60 mV and state c, which also has a low resistance, is reached at still higher potentials

The rate constants k_1, k_2 have the form

$$k_{1,2} = A_{1,2}\, e^{\dfrac{E_{1,2} \pm z\,0.5\,V_M\,23.05}{RT}}\qquad(10)$$

$$E_1 - E_2 = \Delta F\qquad(10a)$$

where $A_{1,2}$ are constants, $E_{1,2}$ are the activation energies for the transition from a to b and b to a, V_M is the membrane potential in volts, 23.05 the conversion factor from electron volts to kilocalories/mole, z the valency of the charge on which V_M acts to cause the configurational change and e and RT have the usual meaning. In the steady state, $dp/dt = 0$ and

$$p = \frac{k_1}{k_1 + k_2}. \tag{11}$$

The same considerations apply to the gates R_2.

Procedure:

1. Calculate p as a function of V_M over the range of ± 160 mV. Using the following values for the other constants: $E_1 = 4$, $E_2 = 1$. $A_{1,2} = 1$, $z = 2$, $e = 2.71$, $RT = 0.6$.

2. Calculate the fraction of the membrane resistance due to gates R_1:

$$R_{M_1} = \frac{R_a R_b}{p(R_a - R_b)\,n + R_b n} \tag{12}$$

for each value of p using $R_a = 10^6$, $R_e = 10^7$, $n = 10^3$.

3. Repeat steps No. 1 and 2 for gates R_2 inverting the voltage sign in Eq. (10).

4. Add R_{M_1} and R_{M_2} to obtain R_M and calculate I_M at each voltage from

$$I_M = \frac{V_M}{R_M}. \tag{13}$$

5. Plot I_M versus V_M.

b) Additional Calculations.

1. Under constant voltage (voltage clamp) the increase of membrane resistance and decrease of current after applied potential steps are monotonic, quasi-exponential functions of time. They can be calculated by integration of Eq. (2) which gives

$$t = -\frac{1}{k_1 + k_2} \ln\left(k_1 - p\,[k_1 + k_2]\right) - t_0 \tag{14}$$

where t_0 is obtained by solving Eq. (11) when $V_M = 0$ [Eq. (10)] and inserting p into Eq. (14).

2. Under constant current conditions, V_M varies with R_M and the rate constants change to

$$k_{1,2} = A_{1,2}\, e^{-\dfrac{E_{1,2} \pm \dfrac{z\,0.5\,V_1\,23.05}{1 + \dfrac{R_i n}{R_a}\left(1 + p\left[\dfrac{R_a - R_b}{R_b}\right]\right)}}{RT}}. \tag{15}$$

In this case, integration of Eq. (9) has to be done by numerical methods. Solutions have produced voltage transitions as in Ref. 7.

3. The resistance decrease at higher voltages (Fig. 5) could be caused by a third voltage dependent resistive state, c, of the gates (Fig. 6).

This assumption introduces two new rate constants for the configurational change from b to c and back and Eq. (9) becomes

$$\frac{dp}{dt} = k_1(1 - [p + q]) + k_4 q - k_2 p - k_3 p \tag{16}$$

$$\frac{dq}{dt} = k_3 p - k_4 q \tag{17}$$

in which q is the fraction of gates in state c, p the fraction in state b, $1 - p + q$ the fraction in state a and k_3, k_4 have the same form as in Eq. (10).

In the steady state: $dp/dt - dq/dt = 0$ and

$$p = \frac{k_1 k_4}{k_1 k_3 + k_2 k_4 + k_1 k_4} \tag{18}$$

$$q = \frac{k_1 k_3}{k_1 k_4 + k_2 k_4 + k_1 k_3} \tag{19}$$

$$R_M = \left(N \left[\frac{1 - (p + q)}{R_a} + \frac{p}{R_b} + \frac{q}{R_c} \right] \right)^{-1} \tag{20}$$

Solutions of Eqs. (18), (19), and (20) allow quantitative calculations of the $I(V)$ curve including the curvature at high values of the membrane potential. They also predict the effect of a resting potential on the $I(V)$ curve which changes its characteristics in one quadrant from negative resistance to rectification. Alternatively, the high voltage resistance decrease could be the result of a direct effect of the high field on the ions as it occurs in free solution (WIEN effect) or in ionic crystals (11). Such effect would make both R_a and R_b of the individual gates voltage dependent, but become pronounced only at higher voltages.

4. Integration of two sets of Eqs. (9) and (12) or (16), (17), and (20) describing the gating of two channel populations with different emf, e.g. cation and anion conducting channels, allows the calculation of action potentials which occur when the ionic conductance of a fraction of the EIM channels is converted by protamine from cationic to anionic. The properties of the experimental system are in agreement with the general requirements of the Hodgkin and Huxley theory; notice that the equivalent circuit diagram (Fig. 6) is very similar to that of HODGKIN and HUXLEY [10]. The theoretical treatment differs because the membrane resistance changes are derived from reaction rate theory instead of empirical permeability functions.

References

1. MUELLER, P., D. O. RUDIN, H. T. TIEN, and W. C. WESCOTT: (1) Nature (Lond.) **194**, 979 (1962). — (2) Symp. Plasma Membrane, N.Y., 1961, Circulation **26** (No. 5, p. 2), 1167 (1962).
2. — — — — J. phys. Chem. **67**, 534 (1963).
3. — — — — In: DANIELLI, J. F., K. G. A. PANKHURST, and A. C. RIDDIFORD, Eds., Progress in surface science **1**, 379. New York: Academic Press 1964.
   SEUFERT, W. D.: Nature (Lond.) **207**, 174 (1965).
4. TIEN, H. T.: J. molec. Biol. **13**, 183 (1965).
5. HANAI, T., D. A. HAYDON, and J. TAYLOR: (1) Proc. Roy. Soc. **281 A**, 377 (1964). — (2) J. gen. Physiol. **48** (No. 5, p. 2), 59 (1965).
6. HUANG, C., and T. E. THOMPSON: J. molec. Biol. **13**, 183 (1965).
   THOMPSON, T. E., and C. HUANG: J. molec. Biol. **16**, 576 (1966).
7. MUELLER, P., and D. O. RUDIN: J. theor. Biol. **4**, 268 (1963).
8. CORKILL, J. M., J. F. GOODMAN, D. R. HAISMAN, and S. P. HARROLD: Trans. Farad. Soc. **57**, 821 (1961).
   PERRIN, J.: Ann. Phys. **10**, 160 (1918).
9. BUKOVSKY, J., and L. KUSHNIR: (Personal communications).
10. HODGKIN, A. L., and A. F. HUXLEY: J. Physiol. (Lond.) **117**, 500 (1952).
11. FRENKEL, J.: Kinetic theory of liquids, p. 40. New York: Dover Publications 1955.

Dissection of Single Nerve Fibres and Measurement of Membrane Potential Changes of Ranvier Nodes by means of the Double Air Gap Method

By R. STÄMPFLI

I. Dissection of a Single Myelinated Nerve Fibre from r. Esculenta [6]

A sciatic gastrocnemius nerve-muscle preparation is obtained in the usual way. It is lifted into air by holding the achilles tendon with a forceps. The glass plate for dissection is then approached from the flat side of the muscle and the preparation brought to rest on it without distorting the nerve. The plate is put on a water-cooled dissection stand. A circular neon-bulb situated underneath provides a dark field illumination.

The dissection is done with fine watchmaker forceps and knee-bent scissors. For the last and finest moments of the dissection sewing needles (No. 11) in needle holders are used. Sharpening of the nedles can be performed on an Arkansas stone if desired. We prefer to renew the needles frequently to avoid a rough surface to which the tissue tends to stick.

After liberating the head of the muscle by cutting the attached blood vessels and after removing them with the adhering connective tissue, one can clearly distinguish the different branches of the sciatic nerve at the level of the knee joint (Fig. 1 A). All the thicker branches are freed from the connective tissue sheath and carefully removed as shown in Fig. 1 B − E. The final dissection is made either on the motor branch going to the head of the muscle or on the sensory branch adhering to it (Fig. 1 D − G). All these steps are commented in the legend of Fig. 1. Two internodes with the three nodes limiting them should be isolated. It is essential to avoid stretching of the fibres. The muscle twitch, mediated by one single fibre, is well seen if the nerve trunk is stimulated with a galvanic forceps. If fibres coming from the motor branch mediate no twitch, they are probably damaged. It is very exceptional to isolate thick sensory fibres, coming from muscle spindles from this motor branch. They can be identified by their shorter action potential and by their low accomodation, which provides a repetitive response to long stimuli.

For the purpose of comparing sensory and motor nerve fibres coming from the same animal, "paired" fibres can be dissected according to Barillot 1966. Transportation of an isolated fibre from the dissection

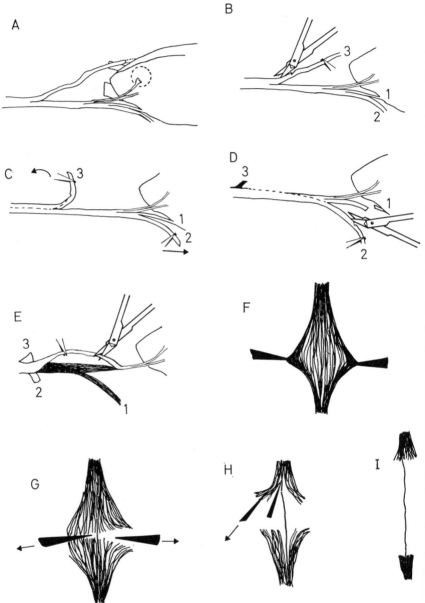

Fig. 1

plate to the recording apparatus is usually done by depositing the non-dissected nerve trunks on the muscle belly and by pulling the fibre alongside the muscle border. By moving the muscle accompanied by a good amount of Ringers solution from the dissection plate into one of the troughs of the recording apparatus, no harm is done to the fibre. Another procedure is given in the contribution of NONNER and STÄMPFLI for small recording chambers, where the nerve trunks have to be short.

II. The Double Air Gap Method [7]

The principle of the method is given in Fig. 2: A single node of Ranvier of an isolated fibre is insulated from its neighbours by keeping it in the axial stream of Ringers or test solution passing a polyethylene tube while the adjacent internodes on either side span air gaps of about 0.5 mm width. The neighbouring nodes and the undissected nerve trunks rest in lateral troughs filled with Ringer's to which 1% procaine has been added. Surface tension and evaporation reduce the external fluid within the air gaps to a film of less than 0.5 μ. The external longitudinal resistance therefore exceeds the longitudinal resistance of the axoplasm by a factor of approximately 10. The fluids of the test compartment and the reference pools on both sides are connected to calomel half cells by saturated KCl-Agar-bridges. The membrane potential changes can be measured by using a feedback amplified connected according to Fig. 3 A. Steady state membrane currents (but not the initial ionic currents) can also be measured by connecting the amplifier in a voltage

---

Fig. 1. Schematic drawing of the isolation of a single myelinated nerve fibre of rana esculenta. A: The head of the gastrocnemius muscle and the main branches of the sciatic nerve at this level. 1 tibial branch containing motor fibres for the distal part of the gastrocnemius. 2 tibial branch following the tibia. 3 peroneal branch. 2 and 3 have to be cut when removing the nerve-muscle preparation from the animal. The small nerve going to the head of the muscle contains two branches, a motor one entering the muscle and a sensory one for the skin area covering the muscle (represented as hatched circle). This branch is disrupted by stripping off the skin. B: The lower blade of the scissors is introduced into an opening of the sheath of the peroneal branch and the sheath is cut open on top of it. C: The peroneal branch is pulled away for the length of opened epineurium. Another forceps holds branch 2 to avoid stretching of the small branches. D: The sheath around branch 2 is now cut open. Branch 1 has been cut previously (twitch). E.: The open sheath is pulled to one side and cut. Higher magnification (about 25 fold). F: The nerve fibres are pulled apart with dissecting needles to form a fan. G: One thick fibre either of the motor or sensory bundle is chosen and all other fibres are cut with the needles. H: The small fibrillae of the endoneural sheath are cut one by one with the needles without stretching the intact fibre. I: Two internodes and three nodes have been isolated and all the other fibres have been cut. The preparation is ready for mounting in a nerve chamber

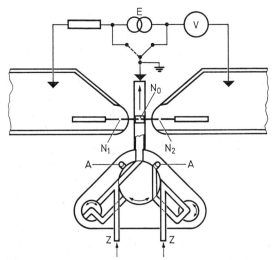

Fig. 2. Schematic drawing of the double air gap method. Node N_1 rests in the axial stream of Ringer's or test solution flowing through the central tube. The lateral nodes are made unexcitable by adding procaine to the Ringer's solution of the neighbouring troughs. The stopcock is designed to allow sudden changes to a new test solution and change of the test solution at will without having remnants of the previous solution in the bore hole of the moving part of the stopcock. The electrical connections are made with calomel half-cells (see Fig. 3)

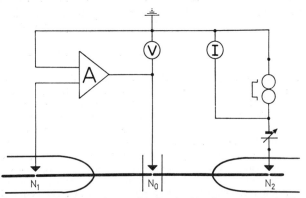

Fig. 3 A. Current and voltage clamp with the double air gap method: The excitable membrane of the central node N_0 can be influenced by constant currents produced by a pulse generator. The potential changes are recorded by a feedback amplifier system (A), which keeps the inside of the nerve fibre at N_0 at constant potential. The feedback output is the absolute value of membrane potential change occurring at the excitable membrane multiplied with the short circuit factor $\left(\dfrac{r_o}{r_o + r_i}\right)$ where r_o = external resistance perunit length, r_i = internal resistance for the same length. There is a variable e.m.f. in series with the pulse generator to provide electrotonic potentials if needed

clamp arrangement, as shown in Fig. 3 B. Typical data of a feedback amplifier to be used in this arrangement are: Input resistance higher than 10^{10} ohms, input capacity less than 10 pF, amplification 10000 up to 30 kC, output impedance less than 100 ohms. The low output resistance gives the possibility to use an ink recorder in parallel with the oscilloscope which is very useful if steady state potentials or membrane currents have to be measured. The change from Ringer's to a test solution is accomplished by a special stopcock within 100 msec or less.

The potential difference obtained by switching to isotonic KCl solution represents 70 to 90% of the value observed with compensating

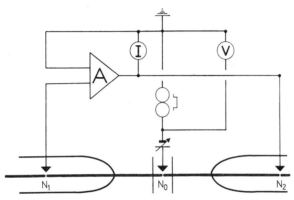

Fig. 3 B. The same elements as in Fig. 3 A can be connected to give a voltage clamp arrangement. The inside at N_0 is still kept at constant potential, this time by passing feedback-current through the axoplasm of the left internode. The potential steps imposed upon the membrane by the pulse generator are thus kept constant. This arrangement is excellent for steady state membrane currents, measured at the end of long impulses. Its time constants are, however, too high to give a reliable voltage clamp during the early permeability changes of the membrane

methods (HUXLEY and STÄMPFLI, 1951). Absolute values of the resting potential of the preparation with respect to zero can be determined by cutting the node in the test compartment with fine scissors at the end of the experiment. Another possibility is to destroy the nodal membrane electrically by excessive hyperpolarization (STÄMPFLI and WILLI, 1957). The values obtained by both methods are in good agreement with the membrane potential differences measured under application of isotonic KCl-solution.

III. Mounting of the Fibre

While the troughs of the nerve chamber (which are movable) are still in contact with the central polyethylene tube, the whole surface of the troughs including the central tube is flooded with Ringer's solution and the preparation is transferred to it by the manipulation described at the end of section A. The muscle is removed and small parts of the nerve trunks are attached to two perspex clamps near the isolated fibre. These attachments can be moved by a rack and pinion arrangement separately or together. The fibre is first extended to its original length, so that it lies in horizontal position in the Ringer's solution covering the chamber. The central node N_0 is then placed exactly on top of the opening of the polyethylene tube. The two troughs are now moved away from the central tube to a distance of approximately 0.5 mm without disrupting the surface of the liquid. The level of the Ringer's solution is now reduced by opening the outflow of the tube by means of a small stopcock. The solution flows off by gravity through a plastic tube. It ends in a cotton wick to avoid the formation of drops and unequal flow of the liquid. Its level can be adjusted before the experiment to permit a constant stream of fluid without overflowing of the opening of the tube. As liquid is leaving the chamber, the bridges of liquid between the lateral pools and the center tubes will disrupt. At this moment the main stopcock is opened and Ringer's solution is allowed to flow along the tube. The air gaps take about 10 min to dry and get a constant high resistance. After this time the experiments can start.

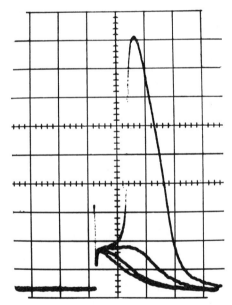

Fig. 4. Action potential at threshold elicited with a pulse of about 20 μsec duration. Superimposed sweeps showing sub-threshold non-linear response. Ordinate: 10 mV Abscissa 0.5 ms per unit

IV. Using the Current Clamp Arrangement

Before turning on the gain of the final stage of the amplifier, the existing small potential difference between the grid pool (right) and the central tube is made zero by shifting the amplifier balance. The connec-

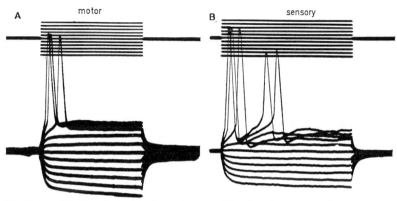

Fig. 5 A and B. Non linear steady state potential values in motor and sensory nerve fibres. A decade of hyperpolarizing and of depolarizing current steps were applied to the nodal membrane and the corresponding membrane potential changes were recorded. Note delayed rectification at end of impulses and repetitive activity in sensory fibre. Abscissa 2 ms. Ordinate 20 mV per unit

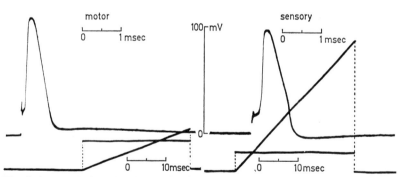

Fig. 6. Accomodation of nodes coming from a motor and from a sensory nerve fibre, measured with a ramp impulse. Upper tracing: Action potential at threshold. Lower tracing: Minimum slope of ramp and rheobasic rectangular impulse superimposed. Accomodation can be given by the minimum rate of rise expressed in rheobases per sec. For the motor fibre this value is more than 200 rheobases/s, for the sensory fibre about 1/4 of this value

tion with the stimulating arrangement is then made. If there is again a small unbalance, the compensating e.m.f. is switched on and adjusted to eliminate it. The gain is turned on and short stimuli are applied. Action potentials can be elicited (Fig. 4). If the stimulating current

11*

is recorded on one beam and the potential change on the other beam of an oscilloscope, the typical non-linearities of the membrane can be observed (Fig. 5). The accommodation can be measured with a ramp impulse in Fig. 6 (Bergman and Stämpfli, 1966). Long impulses tend to induce repetitive activity, particularly in sensory fibres. The refractory period can be demonstrated and the change of threshold can be recorded (Fig. 7).

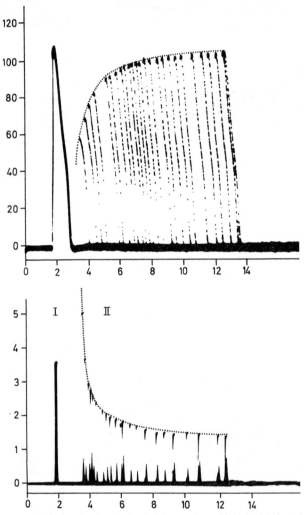

Fig. 7. Upper tracing: Action potentials elicited with a second stimulus within the relative refractory period. Ordinate: membrane potential in mV (corrected for short circuit factor). Lower tracing: Treshold values determined for the same node (ordinate 10^{-10} A per unit). Abscissa for both recordings: time in ms

All these phenomena and many others, such as inactivation, anodal break excitation, the influence of electrotonus etc. can be recorded. A storage scope is particularly useful for such demonstrations.

V. Voltage Clamp Experiments

With this arrangement voltage clamps are possible during a steady state. Neither the amplifier nor the nerve chamber would allow sufficient time resolution to measure the initial sodium currents at room tempera-

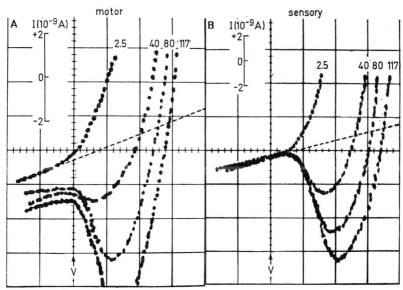

Fig. 8. Steady state current-voltage curves of a Ranvier node in Ringer's solution (2.5 mM/l K) and in high potassium concentrations (40, 80, 117 mM/l) for a motor fibre (A) and a sensory fibre (B). The cathode-ray was modulated to print the current only at the end of a 30 ms potential step. Dotted line = leak current. X-Y storage oscilloscope. Ordinates: total membrane current. Abscissae: membrane potential, 20 mV per division. V resting potential. Note N shaped curve due to shift of V_K to values below V and to potential dependant increase of P_K, providing inward current of potassium between V and V_K

ture. But the arrangement is well suited for recording the membrane currents in high potassium concentrations and the N shaped current voltage curves (Fig. 8). If fast phenomena have to be studied, one should use the method described by NONNER (see page 171). The double air gap method is particularly useful for beginners and for those who want to demonstrate the main electrophysiological elements to the student. It can also give qualitative data of drug actions on Ranvier nodes. Wherever quantitative data are wanted, the method described on page 171 is much more reliable.

References

1. Barillot, J.-Cl.: Etude électrophysiologique comparée des fibres nerveuses myélinisées motrices et sensitives chez Rana esculenta et Xenopus laevis. Thèse du 3ème Cycle Poitiers (1966).

2. Bergman, C., et R. Stämpfli: Différence de perméabilité des fibres nerveuses myélinisées sensorielles et motrices à l'ion potassium. Helv. physiol. pharmacol. Acta 24, 247—258 (1966).

3. Huxley, A. F., and R. Stämpfli: Effect of potassium and sodium on resting and action potentials of single myelinated nerve fibres. J. Physiol. (Lond.) 112, 476—495 (1951).

4. Nonner, W.: A new voltage clamp method for ranvier nodes. Pflügers Arch. (In preparation to appear in 1969).

5. —, and R. Stämpfli: A new voltage clamp method, p. 171. (This volume.)

6. Stämpfli, R.: Bau und Funktion isolierter markhaltiger Nervenfasern. Ergebn. Physiol. 47, 70—165 (1952).

7. — Nouvelle méthode pour enregistrer le potentiel d'action d'un seul étranglement de Ranvier et sa modification par un brusque changement de la concentration ionique du milieu extérieur. J. Physiol. (Paris) 48, 710—714 (1956).

Ionic Currents in the Myelinated Nerve Fibre

By B. FRANKENHAEUSER

A short description will be given of the voltage clamp experiments on the myelinated nerve fibre. This technique allows measurements of the membrane currents associated to step changes of the membrane potential. An analysis of the membrane currents, with the fibre in solutions of various compositions, shows that the ionic currents during step polarizations are passive currents; i.e. the ions move as charged particles in free diffusion in an electric field. The ionic currents can therefore be described by the membrane permeability and the ionic concentrations both sides the membrane. The sodium permeability and the potassium permeability of the membrane depend on membrane potential and on time. These specific permeability changes have been described in a quantitative form. A solution of the equations describing the voltage clamp currents predicts an action potential very similar to the action potential recorded from the myelinated nerve fibre. The analysis follows the main lines of the squid fibre voltage clamp analysis made by HODGKIN and HUXLEY. A reference list is given at the end of this communication covering the major bulk of the original papers on the problem.

The dissected single myelinated fibre was mounted in a recording cell through four pools of Ringer's solution and three petroleum jelly seals at two neighbouring internodes isolating the pools from each other. The node under investigation was located in one pool. Recording and current carrying electrodes were placed in the pools. The membrane potential was controlled by a feed-back arrangement consisting of two feed-back amplifiers. The input of the first amplifier was the potential across one of the seals caused by the longitudinal current. The output of this amplifier was applied as negative series feed-back across the second seal at the same internode. At large amplification this output approximates the membrane potential changes at the node under investigation. The output of this first amplifier was connected to the input of the second feed-back amplifier, while the output from the second amplifier was applied to the other end-pool. The whole system forms negative parallel feed-back which controlls the membrane potential of the node under investigation. Rectangular pulses of current were applied to the input of the second feed-back amplifier. This changes the membrane potential in desired rectangular steps. The membrane current associated with these step changes of potential were measured.

A step depolarization was associated with : (a) A short-lasting sudden surge of capacity current (I_C); (b) A leak current (I_L) which was more or less ohmic; (c) An initial transient current, reversing its direction at a definite potential. Since the point of reversal depended on sodium concentration as the sodium equilibrium potential the initial current is held to be carried mainly by sodium ions (I_{Na}); (d) A delayed current mainly carried by potassium (I_K); (e) A delayed current (I_p) through a permeability channel with changing but relatively unspecific permeability.

These currents were analysed in detail and empirical equations were fitted to the experimentally measured and separated currents. The measurements indicate that the permeabilities are largely independent of concentrations while the specific ionic currents depend on the concentration of the same ions as predicted by the constant field equation but are largely independent of the concentrations of the other ions. Calcium and magnesium, however, change the specific permeabilities.

It is evident that during a voltage clamp step the solutions of these differential equations are independent of each other. If the currents change the membrane potential as in a not clamped nerve, then the solution of one equation depends on the other ones. The equation system can then be solved with as an example the Runge-Kutta method. If the analysis is correct and covers all the major currents then such solutions ought to predict an action potential and excitability properties which reasonably agree with the properties of the nerve. Such numerical computations were made and they indicated a quite good agreement between the predictions and the response of the nerve.

It may be concluded that the voltage clamp analysis accounts in a satisfactory way for the excitability properties of the nerve. The analysis gives no account for the mechanism of the permeability changes.

This investigation has been supported by the Swedish Medical Research Council (Project No. 14X-545) and Stiftelsen Therese och Johan Anderssons Minne.

References

Bennet, M. R.: An analysis of the surface fixed-charge theory of the squid giant axon membrane. Biophys. J. 7, 151—164 (1967).

Cole, K. S., and J. W. Moore: Potassium ion current in the squid giant axon: dynamic characteristic. Biophys. J. 1, 1—14 (1960).

Dodge, F. A.: Ionic permeability changes underlying nerve excitation. Biophysics of physiological and pharmacological actions, p. 112—143. Washington: Amer. Ass. Adv. Sci. 1961.

—, and B. Frankenhaeuser: Membrane currents in isolated frog nerve fibre under voltage clamp conditions. J. Physiol. (Lond.) 143, 76—90 (1958).

— — Sodium currents in the myelinated nerve fibre of Xenopus leavis investigated with the voltage clamp technique. J. Physiol. (Lond.) 148, 188—200 (1959).

FRANKENHAEUSER, B.: The hypothesis of saltatory conduction. Cold Spr. Harb. Symp. quant. Biol. **17**, 27—32 (1952).
— (1) A method for recording resting and action potentials in the isolated myelinated frog nerve fibre. J. Physiol. (Lond.) **135**, 550—559 (1957).
— (2) The effect of calcium on the myelinated nerve fibre. J. Physiol. (Lond.) **137**, 245—260 (1957).
— Steady state inactivation of sodium permeability in myelinated nerve fibres of *Xenopus laevis*. J. Physiol. (Lond.) **148**, 671—676 (1959).
— (1) Quantitative description of sodium currents in myelinated nerve fibres of *Xenopus laevis*. J. Physiol. (Lond.) **151**, 491—501 (1960).
— (2) Sodium permeability in toad nerve and in squid nerve. J. Physiol. (Lond.) **152**, 159—166 (1960).
— (1) Delayed currents in myelinated nerve fibres of *Xenopus laevis* investigated with voltage clamp technique. J. Physiol. (Lond.) **160**, 40—45 (1962).
— (2) Instantaneous potassium currents in myelinated nerve fibres of *Xenopus laevis*. J. Physiol. (Lond.) **160**, 46—53 (1962).
— (3) Potassium permeability in myelinated nerve fibres of *Xenopus laevis*. J. Physiol. (Lond.) **160**, 54—61 (1962).
— (1) A quantitative description of potassium currents in myelinated nerve fibres of *Xenopus laevis*. J. Physiol. (Lond.) **169**, 424—430 (1963).
— (2) Inactivation of the sodium-carrying mechanism in myelinated nerve fibres of *Xenopus laevis*. J. Physiol. (Lond.) **169**, 445—451 (1963).
— Computed action potential in nerve from *Xenopus laevis*. J. Physiol. (Lond.) **180**, 780—787 (1965).
—, and A. L. HODGKIN: The action of calcium on the electrical properties of squid axons. J. Physiol. (Lond.) **137**, 218—244 (1957).
—, and A. F. HUXLEY: The action potential in the myelinated nerve fibre of *Xenopus laevis* as computed on the basis of voltage clamp data. J. Physiol. (Lond.) **171**, 302—315 (1964).
—, and L. E. MOORE: (1) The effect of temperature on the sodium and potassium permeability changes in myelinated nerve fibres of *Xenopus laevis*. J. Physiol. (Lond.) **169**, 431—437 (1963).
— — (2) The specificity of the initial current in myelinated nerve fibres of *Xenopus laevis*. Voltage clamp experiments. J. Physiol. (Lond.) **169**, 438—444 (1963).
—, and A. B. VALLBO: Accommodation in myelinated nerve fibres of *Xenopus laevis* as computed on the basis of voltage clamp data. Acta physiol. scand. **63**, 1—20 (1965).

HILLE, B.: Common mode of action of three agents that decrease the transient change in sodium permeability in nerves. Nature (Lond.) **210**, 1220—1222 (1966).

HODGKIN, A. L., and A. F. HUXLEY: (1) Currents carried by sodium and potassium ions through the membrane of the giant axon of *Loligo*. J. Physiol. (Lond.) **116**, 449—472 (1952).
— — (2) The components of membrane conductance in the giant axon of *Loligo*. J. Physiol. (Lond.) **116**, 473—496 (1952).
— — (3) The dual effect of membrane potential on sodium conductance in the giant axon of *Loligo*. J. Physiol. (Lond.) **116**, 497—506 (1952).
— — (4) A quantitative description of membrane current and its application to conduction and excitation in nerve. J. Physiol. (Lond.) **117**, 500—544 (1952).
— —, and B. KATZ: Measurement of current-voltage relations in the membrane of the giant axons of *Loligo*. J. Physiol. (Lond.) **116**, 424—448 (1952).
—, and B. KATZ: The effect of sodium ions on the electrical activity of the giant axon of the squid. J. Physiol. (Lond.) **108**, 37—77 (1949).

Huxley, A. F., and R. Stämpfli: Evidence for saltatory conduction in peripheral myelinated nerve fibres. J. Physiol. (Lond.) **108**, 315—339 (1949).

— — Direct determination of membrane resting potential and action potential in single myelinated nerve fibres. J. Physiol. (Lond.) **112**, 476—495 (1951).

Koppenhöfer, E.: Die Wirkung von Tetraäthylammoniumchlorid auf die Membranströme Ranvierscher Schnürringe von *Xenopus laevis*. Pflügers Arch. ges. Physiol. **293**, 34—55 (1967).

Narahashi, T., N. C. Anderson, and J. W. Moore: Tetrodotoxin does not block excitation from inside the nerve membrane. Science **153**, 765—767 (1966).

—, J. W. Moore, and W. R. Scott: Tetrodotoxin blockage of sodium conductance increase in lobster giant axons. J. gen. Physiol. **47**, 965—974 (1964).

Stämpfli, R.: Bau und Funktion isolierter markhaltiger Nervenfasern. Ergebn. Physiol. **47**, 69—165 (1952).

— Conduction and transmission in the nervous system. Ann. Rev. Physiol. **25**, 493—522 (1963).

Tasaki, I.: Nervous Transmission. Springfield: Thomas 1953.

Vallbo, A. B.: Accommodation related to inactivation of the sodium permeability in single myelinated nerve fibres from *Xenopus laevis*. Acta physiol. scand. **61**, 429—444 (1964).

A New Voltage Clamp Method

By **W. Nonner** and **R. Stämpfli**

The preceding contribution has outlined the principles of measuring ionic currents by means of voltage clamp in myelinated nerve fibres. The method (Fig. 1) developed by Frankenhaeuser has been demon-

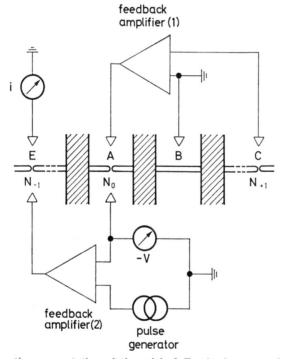

Fig. 1. Schematic representation of the original Frankenhaeuser voltage clamp. ABCE = compartments of the nerve chamber, separated by vaseline seals (hatched areas). The first feedback amplifier measures membrane potential of node N_0. The second amplifier clamps to the potential imposed by the pulse generator

strated to the participants of the course. At this time, Nonner had just started using a new method which since then has proved its superiority. It seems therefore justified to describe it shortly and to give a few examples of the results obtained on rat nerve fibres, illustrating also the general outline given by Frankenhaeuser in the preceding contribution.

The new voltage clamp method is represented in Fig. 2. It differs from the old one as follows:

The resistance of the external pathway for detecting a current flow between N_0 and N^{+1} is strongly increased by introducing an air gap. This improves the sensitivity of the potential measuring feedback system sufficiently to use it directly for clamping the membrane of N_0, without making use of a second amplifier. Drift is reduced and the "attenuation

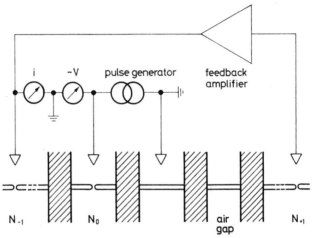

Fig. 2. New voltage clamp: The only feedback amplifier keeps the potential inside N_0 constant. Changes of outside potential are imposed by pulse generator. Note air gap to increase input resistance

artifact" (Dodge and Frankenhaeuser, 1958) is abolished. The use of one single amplifier simplifies the technique and improves the frequency response. A theoretical treatment of this circuit is given by Nonner (1969).

The nerve chamber is made of two movable troughs on either side of a perspex partition (second seal from the right in Fig. 2). A smaller, movable trough which can be adjusted in its position to give an optimum width to the air gap (between 120 and 180 μ) is fitted into the right hand trough. The surfaces of the 100 μ wide ridges, forming the borders of the troughs and of the partition mentioned afore are coated with a very thin layer of silicone grease. The whole chamber is then flooded with Ringer's and remaining air bubbles are removed with a fine pipette.

The nerve fibre is dissected in the usual way (page 158) on a glass plate. The undissected nerve trunks on either side of the two isolated internodes are cut to approximately 5 mm length. By means of a forceps, an opening is made in the large proximal trunk and the small distal one is then pulled into it to permit easy transportation of the fibre without pulling it, simply by moving the proximal trunk with the forceps within the fluid, covering the glass plate. This is now lifted to the level of fluid protruding from the chamber and the two fluids are made to join. At this moment, the preparation is rapidly moved into one of the troughs and the fluids are separated again by removing the glass plate. The trunks are pulled apart with dissection needles and fixed horizontally to two movable forceps, mounted on a micromanipulator. The fibre is stretched to its original length within the fluid above the ridges and adjusted to

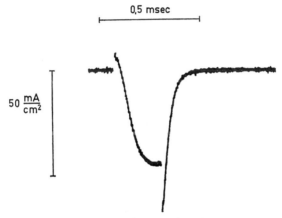

Fig. 3. Membrane current of rana esculenta node at room temperature during a voltage step of 46 mV interrupted at maximum inward current. Note tail current (time resolution is about 10 μsec)

have node N_0 exactly on top of the test gap. The fibre is then lowered onto the ridges, the trunks are freed from the forceps and fixed in the bottom of the external troughs with vaseline sausages squeezed out through a hypodermic needle from a syringe. With this same method, vaseline sausages of 100 μ diameter are deposited on all ridges, forming the future seals and fixing the fibre in its position. The chamber is now connected to calomel half cells by isotonic KCl-bridges and the test compartment to a stopcock for circulating either Ringer's or test solutions through the test compartment. The outflow of circulating fluid is made by a vertical tube, adjusted to the appropriate level and connected to a suction-pump. The level of the fluid is now slowly lowered until the vaseline seals emerge. The fluid within the air gap is removed by sudden suction with a small pipette. The nerve fibre is cut in the right hand trough to avoid the flow of feedback current across the non-linear resistance of N_{-1}. Another possibility to avoid changes of the series resistance of the current path is to treat N_{-1} with 1% formaldehyde solution for 30 sec and to fill the trough with isotonic KCl afterwards (Hille, 1967).

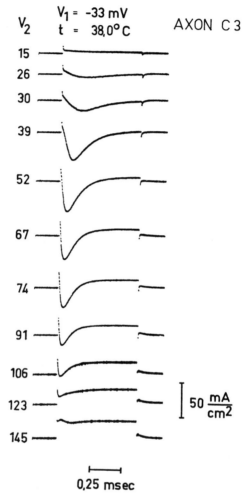

Fig. 4. Typical voltage clamp run of rat node at 38 °C. Figures on left indicate step amplitude. V_1 is a hyperpolarizing step at 50 ms duration

The node is now ready for voltage or current clamps. The amplifier contains the necessary compensating circuits to bring the potential difference between the input and the ground compartment to zero. A small hyperpolarizing pulse is now applied to the test compartment and the amplifier gain opened at high current sensitivity of the oscilloscope.

Other built in compensating circuits can now be adjusted to give a maximum time resolution between ionic and capacitive currents at the beginning and end of the impulse. The best check of the amplifier matching is to produce a maximum initial inward current by a short depolarizing impulse and to observe the tail sodium current after the impulse. Fig. 3 gives a typical example of the time resolution obtained. The arrangement is now ready to give voltage clamp measurements for one to several hours without further major adjustment. (The preparation can also be used in a current clamps circuit, by inverting the connections to A and E.) A typical run for initial currents is given in Fig. 4 and the current voltage curve derived from it in Fig. 5. These results were obtained from a nodal membrane of a sensory rat nerve fibre at 38 °C. It would have been im-

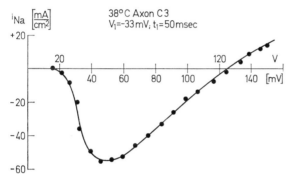

Fig. 5. Peak sodium currents plotted from Fig. 4. Note V_{Na} at 128 mV

possible to get them with the FRANKENHAEUSER arrangement which already gave difficulties to measure initial currents at room temperature if not particularly thick motor fibres of Xenopus laevis were used.

The feedback amplifier was developed by physicist D. WEYMANN. It is a fully transistorized and chopper-stabilized broad-band DC-amplifier. Maximum amplification 90 db, transit frequency $4 \cdot 10^{10}$ C/s, input impedance 470 Megohms 7 pF, output impedance 50 ohms, grid current less than 10^{-13} Amps, DC input level adjustable by ± 120 mV.

References

DODGE, F. A., and B. FRANKENHAEUSER: Membrane currents in isolated frog nerve fibre under voltage clamp conditions. J. Physiol. (Lond.) 143, 76—90 (1958).

HILLE, B.: A pharmacological analysis of the ionic channels of nerve. Thesis, The Rockefeller University, 1967.

NONNER, W.: A new voltage clamp method for ranvier nodes. Pflügers Arch. (In preparation, to appear in 1969.)

Electrical Activity of Synapses.
The Crayfish Neuromuscular Junction

By J. Dudel

I. Introduction

a) **Technical Problem.** Measurement of excitatory and inhibitory postsynaptic potentials, postsynaptic current and presynaptic activity.

b) **Principle of Method.** A suitable muscle has to be isolated. The excitatory and the inhibitory axons leading to the muscle have to be prepared for separate stimulation. The excitatory and the inhibitory postsynaptic potentials are recorded using intracellular microelectrodes [Fatt and Katz, 1953 (1, 2)]. Postsynaptic activity from single junctions is recorded by means of extracellular microelectrodes [Del Castillo and Katz, 1956; Dudel and Kuffler, 1961 (1, 3)]. Postsynaptic membrane conductance is determined by application of current through a second intracellular microelectrode. By this method also the nature of the postsynaptic inhibitory potential can be shown by demonstrating its reversal potential and the dependence of the reversal potential on extracellular chloride concentration [Fatt and Katz, 1953 (2); Boistel and Fatt, 1958; Grundfest, Reuben and Rickles, 1959; Dudel and Kuffler, 1961 (3)]. Presynaptic activity of the nerve terminals is recorded by means of extracellular microelectrodes [Dudel, 1963, 1965 (1)].

c) **Special Applications.** Demonstration of quantal release of excitatory transmitter [Dudel and Kuffler, 1961 (1)]. Demonstration of synaptic facilitation at increased rates of stimulation [Katz and Kuffler, 1946; Dudel and Kuffler, 1961 (2); Dudel, 1965 (1)].

Differentiation of post- and presynaptic inhibition showing the reduction of postsynaptic excitatory current during presynaptic inhibition [Dudel and Kuffler, 1961 (3)]. Presynaptic inhibition can also be shown to affect potential changes generated in the excitatory nerve terminals [Dudel, 1965 (3)].

Determination of post- and presynaptic inhibitory effects of γ-aminobutyric acid and related drugs [Robbins, 1959; Dudel and Kuffler, 1961 (3); Dudel, 1965 (2, 4)].

II. Instruments and Solutions

A conventional setup for peripheral neurophysiology is needed.

a) Recording of Potentials. Potentials are recorded by means of glass-capillary microelectrodes. Electrodes for intracellular penetration have tip diameters of 1 μ or less, they are filled with a 3 molar solution of KCl; if they are used for passing current, they are filled with 2 M/l of potassium citrate. Electrodes used for extracellular recording have tip diameters of 1 — 2 μ, they are filled with nearly saturated NaCl solution. Their electrical resistance is 1 — 4 MΩ. For the insertion of the microelectrodes two micromanipulators (LEITZ) are used. For intracellular recording any stable cathode follower serves as input stage, modifications of the circuit described by ELUL and TAMARI (1963) or electrometer tubes in combination with operational amplifiers (see TRAUTWEIN et al., 1965) are used. For extracellular recording the input stage should have a low noise level. Simple cathode followers consisting of a low noise triode are found most useful. Very low level extracellular records can be discerned from noise larger than the signal using a computer for averaging repetitive signals. A computer of average transients (CAT 400 B, Technical Measurements Corporation) was employed [see DUDEL, 1965 (1)].

The signals are amplified using conventional dc- or low level ac-amplifiers. The signals are displayed on an oscilloscope. A true two beam instrument (e.g. TEKTRONIX 565) is to be preferred; this allows simultaneous display of the signal on a fast and a slow time base.

b) Stimulators. For stimulation of the axons stimuli of about 1 V lasting 0.5 ms are necessary, they should be isolated from ground in order to minimize artefacts. GRASS S5 stimulators are used. The axons are stimulated either at a constant rate of 1 — 20 /s or in trains of stimuli. The trains are obtained by gating the nerve-stimulator for the desired periods of stimulation of 0.1 to 1 s. TEKTRONIX Type 161 pulse units supply the gating voltage. The latter type of stimulator also is used to generate the rectangular pulses through the current electrode.

c) Solutions. The preparations are kept in a modified VAN HARRE-VELD (1936) solution, containing in mM/l: NaCl 195, KCl 5.4, CaCl$_2$ 13.5, MgCl$_2$ 2.6 and tris-maleate-buffer 10 (pH 7.5). In solutions with raised KCl concentrations the NaCl concentration is reduced by an equivalent amount.

III. Execution

Preparation of the crayfish opener of the claw muscle.

Crayfish of the species Astacus fluviatilis (Europe) or Orconectes virilis (North America) are used. The m. abductor of the dactyl from the first pair of walking legs was found most useful. The leg is cut off at the

natural point of autotomy between the ischio- and the meropodite. The
ligaments and hinges between meropodite and carpopodite are cut and
meropodite and carpopodite are pulled apart gently. By this procedure
the nerve bundles lying in the meropodite are exposed. The carpopodite
then is fixed in a perspex block, the exposed nerve bundles lying in a
pool of saline (see Fig. 1).

The opener muscle is innervated by two nerve fibers, an excitatory
and an inhibitory axon. These nerve fibers lie in different nerve bundles
in the meropodite. The respective nerve bundles are identified by electrical stimulation. If the excitatory axon is contained in the nerve bundle,
the 'claw' opens on stimulation. The inhibitory axon to the opener
muscle always lies very close to the excitatory axons leading to the anta-
gonist of the opener muscle, the adductor of the dactylopodite. The

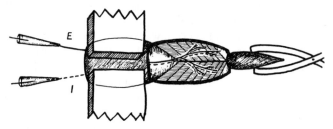

Fig. 1. Scheme of the preparation of the abductor of the dactyl in the crayfish,
viewed from above. Adductor muscle removed. E = excitatory axon, I = inhibitory
axon

inhibitory axon to the opener therefore can be found eliciting a contrac-
tion in the adductor. The nerve bundles containing the excitatory and
the inhibitory axon then are sucked into fluid electrodes (see Fig. 1).
These consist of glass tubes with a small opening, large enough for a
nerve bundle to go through, connected to a syringe by fine tubing.
Various lengths of nerve can be pulled up into the tube, together with
physiological solution. The stimulus is then applied between the inside
and the outside of the glass tube.

The carpopodite is fixed with the ventral side facing upward, so that
the moveable dactyl lies at the bottom of the chamber. The opener
muscle is exposed by removing the upper portion of the propodite together with the m. adductor of the dactyl which inserts there. The surface
of the opener muscle is cleaned of sensory nerve bundles, arteries and
connective tissue. The dactyl is fixed with a metal clip (see Fig. 1). The
muscle is viewed with a dissecting microscope; in dark field illumination the course of the nerve fibers can be seen. The muscle is relatively
thin, consisting of several layers of short fibers which originate on the

exosceleton and are inserted on a central tendon. Most of the fibers which were used were 200 to 300 μ in diameter and 2 – 3 mm long. The intact exosceleton around the muscle fibers forms a natural chamber with a volume of not more than 0.1 ml.

The experiments.

Intracellular recording: A microelectrode is inserted into a muscle fiber. On stimulation of the excitatory or/and the inhibitory nerve fiber (see Fig. 2) postsynaptic excitatory (e.p.s.p.s) and inhibitory (i.p.s.p.s) potentials and their interactions are observed. The membrane resistance changes during postsynaptic potentials are determined by observing their effect on electrotonic potentials, which are generated by current

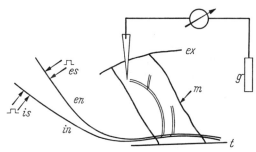

Fig. 2. Schematic view of the preparation. *m* muscle fiber; *ex* exosceleton; *t* tendon; *en* excitatory nerve fiber; *in* inhibitory nerve fiber; *es* excitatory stimulation electrodes; *is* inhibitory stimulation electrodes; *g* ground

pulses applied through a second intracellular electrode. The effects of an increased extracellular K^+ concentrations or of inhibitory drugs like γ-amino-butyric acid are seen after dropping the respective solutions on the muscle preparation.

Extracellular recording: When an extracellular microelectrode is placed on the muscle surface, no signal is seen as a rule after stimulation of the excitatory nerve. If, however, a thorough search is conducted along the fiber, making on the average 30 – 50 contacts, usually a spot is found where potentials appear with nerve stimulation. These spots are sharply localized and the potentials disappear with a lateral movement of the electrode tip of several microns.

The extracellularly recorded e.p.s.p.s originate at a single synaptic region. Their quantal composition can be analysed. The extracellularly recorded e.p.s.p.s can also be inhibited by stimulation of the inhibitory fiber. This inhibition is presynaptic.

At the synaptic spot the extracellularly recorded e.p.s.p. is preceded by a potential change originating in the nerve terminal. The influence of

12*

frequency of excitatory stimulation, inhibitory stimulation, etc. on this nerve terminal potential can be studied.

References

Boistel, J., and F. Fatt: Membrane permeability change during inhibitory transmitter action in crustacean muscle. J. Physiol. (Lond.) 144, 176—191 (1958).

Del Castillo, J., and B. Katz: Localization of active spots within the neuromuscular junction of the frog. J. Physiol. (Lond.) 132, 630—649 (1956).

Dudel, J.: Presynaptic inhibition of the excitatory nerve terminal in the neuromuscular junction of crayfish. Pflügers Arch. ges. Physiol. 277, 537—557 (1963).

— (1) Potential changes in the crayfish motor nerve terminal during repetitive stimulation. Pflügers Arch. ges. Physiol. 282, 323—337 (1965).

— (2) Presynaptic and postsynaptic effects of inhibitory drugs on the crayfish neuromuscular junction. Pflügers Arch. ges. Physiol. 283, 104—118 (1965).

— (3) The mechanism of presynaptic inhibition at the crayfish neuromuscular junction. Pflügers Arch. ges. Physiol. 284, 66—80 (1965).

— (4) The action of inhibitory drugs on nerve terminals in crayfish muscle. Pflügers Arch. ges. Physiol. 284, 81—94 (1965).

—, and S. W. Kuffler: (1) The quantal nature of transmission and spontaneous miniature potentials at the crayfish neuromuscular junction. J. Physiol. (Lond.) 155, 514—529 (1961).

— — (2) Mechanism of facilitation at the crayfish neuromuscular junction. J. Physiol. (Lond.) 155, 530—542 (1961).

— — (3) Presynaptic inhibition at the crayfish neuromuscular junction. J. Physiol. (Lond.) 155, 543—562 (1961).

Elul, R., and A. Tamari: An amplifier with constant unity gain for microelectrode studies. Electroenceph. clin. Neurophysiol. 15, 118—122 (1963).

Fatt, P., and B. Katz: (1) Distributed 'end-plate potentials' of crustacean muscle fibres. J. exp. Biol. 30, 433—439 (1953).

— — (2) The effect of inhibitory nerve impulses on a crustacean muscle fibre. J. Physiol. (Lond.) 121, 374—389 (1953).

Grundfest, H., J. P. Reuben, and W. H. Rickles: The electrophysiology and pharmacology of lobster neuromuscular synapses. J. gen. Physiol. 42, 1301—1323 (1959).

Harreveld, A. von: Physiological solution for freshwater crustaceans. Proc. Soc. exp. Biol. (N.Y.) 34, 428—432 (1936).

Katz, B., and S. W. Kuffler: Excitation of the nerve-muscle system in crustacea. Proc. roy. Soc. B. 133, 374—389 (1946).

Robbins, J.: The excitation and inhibition of crustacean muscle by amino acids. J. Physiol. (Lond.) 148, 39—50 (1959).

Trautwein, W., J. Dudel, and K. Peper: Stationary S-shaped current voltage relation and hysteresis in heart muscle fibers. Excitatory phenomena in Na^+-free bathing solutions. J. cell. comp. Physiol. Suppl. 2, 66, 79—96 (1965).

Appendix
Double Tracer Techniques

By E. OBERHAUSEN and H. MUTH

Double tracer techniques have first been employed in studies of metabolic reactions. By labelling one molecule with two different radio-nuclides it was possible to follow the distribution and chemical modification of many metabolic substrates in whole animals, tissues, and cells. In permeability research the simultaneous application of several isotopes in one experiment is quite often advantageous or necessary. For example, the measurement of the fluxes of the different alkali ions under strictly identical conditions requires the simultaneous determination of the radio-isotopes of more than one alkali metal species. Since for each alkali metal species a gamma emitting isotope exists and since the energies of these emitters are sufficiently different, such simultaneous determinations can actually be performed. For the simultaneous measurement of ion and water movements, tritiated water can be employed besides the gamma emitting ions. Finally, if one wants to measure the two opposizing unidirectional sodium fluxes in a two compartment system, it is necessary to use two different sodium isotopes.

Many of the radionuclides used in permeability research are 'pure' beta emitters. They emit besides a beta particle a neutrino, but the latter is not detected by the conventional detecting devices usually applied in biological research. The typical pure beta emitters most frequently used in biological investigations include ^{14}C, ^{33}P, ^{35}S, and ^{3}H. There are other beta emitters which emit, in addition to the beta particle and the neutrino, one or several gamma quantums. If one wants to measure the activities of these radionuclides, it is usually preferable to determine the gamma radiation. Typical radionuclides of this category include ^{137}Cs, ^{40}K, and ^{80}Rb. For the present purpose these radionuclides will be called gamma emitters. However, one has to keep in mind that a beta particle is also emitted.

There exists a third category of radionuclides which plays a role in biological research: the positron emitters. The positron travels only a short distance and combines then with an electron. This combination results in the emission of two gamma rays with an energy of 0.51 MeV.

Together with the positron, other gamma rays can be emitted also. A typical positron emitter is ²²Na. The activity of positron emitters can be determined like that of gamma emitters.

One can distinguish the following three cases for double labelling:

1. Labelling with 2 gamma emitters,
2. labelling with 2 beta emitters,
3. labelling with gamma and beta emitter.

If suitable radionuclides are available it is even possible to use three or four tracers in the same experiment and to measure them in the same samples. This can be done without principal changes in the technique of the measurement. However, as will be described below, the systems of two linear equations used for the evaluation of experiments with two tracers has to be enlarged by as many additional equations as additional radionuclides are present.

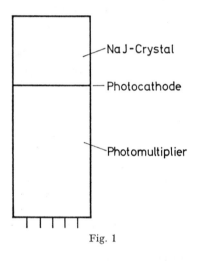

Fig. 1

Before discussing the three types of double labelling mentioned above, the basic principles for measuring gamma and beta radiations should be remembered. The simultaneous use of two tracers requires a measuring device that can distinguish between the two radionuclides. This can be achieved if the output pulse of the measuring instrument is proportional to the energy absorbed in the radiation detector. This condition is met in scintillation counters with inorganic crystals. The two main parts of such a scintillation counter consist in a Na I, Tl activated crystal and a photomultiplier (Fig. 1). When a gamma ray is absorbed in the crystal, photons of visible or near ultra violett light are created. The number of photons is proportional to the absorbed energy of the gamma ray. The photons fall on the photocathode of the photomultiplier and thereby liberate electrons. In the dynode system of the multiplier these electrons are multiplied about 10^6 fold. At the anode of the multiplier one therefore obtains an electrical impulse which is proportional to the absorbed energy of the gamma ray. The distribution of the pulse heights at the anode for a certain gamma emitting radionuclide is shown in Fig. 2. The gamma ray emitter used, was ¹³⁷Cs which emitts gamma rays of an energy of 0.66 MeV. Since the gamma rays from ¹³⁷Cs are monoenergetic one would expect a single line of the corresponding energy. However, instead of a single line there is a maximum with a definite width. This is caused by

the fact that both, the creation of photons and electrons and the multiplication of electrons are statistical processes. The hights of the pulses
produced by single events are therefore unequal and statistically distributed around the most probable value. The width of the distribution
curve depends also on the quality of the crystal used.

The resolution of a scintillation counter is defined as the width at half
height of the distribution curve expressed in percent of the gamma energy.
The resolution thus defined depends on the energy of the gamma radiation. It is therefore normally measured with a specific isotope, ^{137}Cs. With
good scintillation counters one attains a resolution of better than 10%.
If several radionuclides are to be determined simultaneously, resolution
is a very important factor. This is especially true if the energies of the

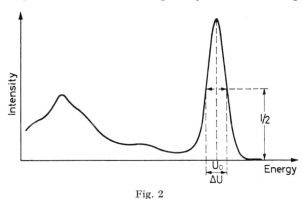

Fig. 2

different gamma emitters are of similar magnitude. From Fig. 2 it can be
seen further that besides the maximum there exists a continuum of
smaller pulse heights. This continuum is caused by those gamma rays
which interact with the crystal in a Compton effect. In a Compton effect
only part of the energy of the gamma ray is absorbed. So this small
pulses represent that part of the energy which is absorbed in a Compton-
effect whereas the higher pulse around the maximum represent the total
energy which is absorbed in a photo effect.

To record a gamma spectrum a few other apparatus besides the
scintillation counter are necessary. This is shown in Fig. 3. The output of
the photomultiplier is lead to a preamplifier which is in most cases only
an impedance transformer. Then follows a linear amplifier with a gain
of about 1000 to amplify the pulses to a pulse height of several volts. The
registration of the pulse height spectrum can be achieved with a single
channel pulse height analyser and a scaler or with a multichannel
analyser. In the single channel analyser only those pulses pass through
which have a pulse height between h and $h + \Delta h$. To register the whole

spectrum it is necessary to change successively the value of h from near 0 to the highest pulse heights. Thus a great number of measurements is necessary to obtain the whole spectrum. In a multichannel analyser all the incoming pulses are stored according to their energy in different channels. Thus with one single measurement one obtains the whole gamma spectrum.

Since gamma rays are very penetrating, self absorption within the sample can be neglected. This does not apply to beta rays. Their short

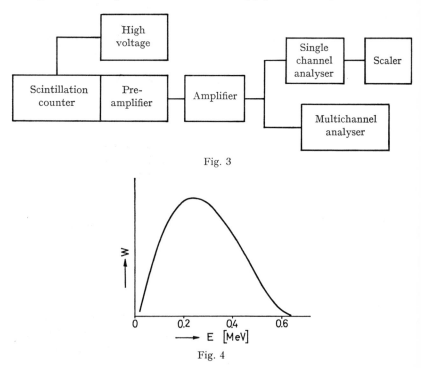

Fig. 3

Fig. 4

range represents the most important difficulty for their measurement. This is especially true for the soft beta rays emitted by ^{14}C, ^{3}H, and ^{35}S. These beta emitters are therefore usually counted with liquid scintillation counters. The radioactive substances are dissolved in a liquid scintillator where no self-absorption occurs since each radioactive atom is surrounded with molecules of the liquid scintillator which emits photons when hit by a beta particle.

The main problems with liquid scintillators are quenching of the fluorescent light and poor miscibility of radioactive solutions with the scintillator fluid.

Another difficulty, especially when two or more beta emitters have to be counted in the same sample, is the energy distribution of the beta rays. Here we have no monoenergetic rays, but a broad continuum which

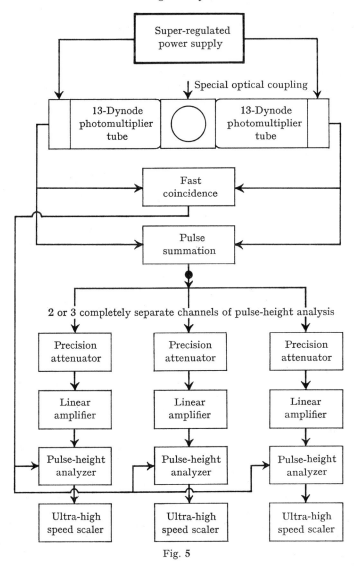

Fig. 5

extends up to the maximum energy of the beta emitter. Such an energy distribution is shown in Fig. 4. Very similar is the pulse hight distribution which one gets from a beta emitter in a liquid scintillator. As there are no defined peaks there is not much use for a multichannel analyser. In liquid scintillation counting one works therefore with analysers having one or several simple channels. A diagram of a complete apparatus is shown in Fig. 5. The pulses from the low energy beta emitters are small; they are even very small as the efficiency for producing photons is smaller in liquids than it is in crystals. The pulse heights resulting from the beta rays are therefore of the same magnitude as those from the thermal emission of the photocathode. This would cause a high background if one would work with a single phototube at room temperature. To reduce the background, the system is mounted in a cooling box and two photomultipliers are working together in an anticoincidence circuit.

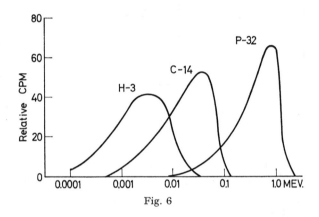

Fig. 6

Thus, only if a pulse occurs on both anodes at the same time are they accepted by the instrument. As the thermal emission is random, it cannot produce pulses on both anodes at the same time. After summation of the pulses they are fed into several linear amplifiers. Behind each amplifier is a single channel analyser. By different settings of the analysers one can measure several beta emitters in the same sample. The pulse height distributions from ^{3}H, ^{14}C, and ^{32}P are shown in Fig. 6.

The measurement of ^{42}K and ^{22}Na may serve as an example of double labelling. Fig. 7 shows the gamma spectra of the two radionuclides. ^{42}K has a maximum at 1.52 MeV and ^{22}Na has two maxima at 0.51 and 1.28 MeV. The first maximum of ^{22}Na is caused by recombination of the emitted positrons with electrons. If both radionuclides are in the same sample the measurement has to be made in two energy channels which are chosen in such a way that the peaks from the different isotopes fall into different channels. The channel should not be too narrow so that the steep sides of the peaks do not coincide with the limit of the channel. Small fluctuations in amplification could then cause no great changes in counting efficiency. This precaution is very important if one uses single channel analysers. With multichannel analysers such fluctuations will be seen from the shape of the spectrum and can be considered in evaluating the results. For the measurement of ^{22}Na one can choose between two peaks. In most cases it is better to use the higher energy peak, since at higher energies background is smaller. In measuring ^{22}Na

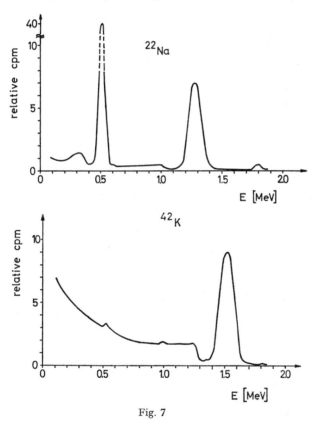

Fig. 7

and ^{42}K a good solution is therefore to work with a first channel from 1.1 to 1.4 and a second from 1.4 to 1.7 MeV. After measurement the two activities can be calculated by means of the following linear equations:

$$I_1 = C_{11} \cdot \text{Na} + C_{12} \cdot \text{K} \qquad (1)$$
$$I_2 = C_{21} \cdot \text{Na} + C_{22} \cdot \text{K} .$$

The symbols means:

I_1 = c.p.m. in channel 1,
I_2 = c.p.m. in channel 2,

Na = activity of ^{22}Na in the sample,
K = activity of ^{42}K in the sample,
C_{11} = calibration constant for ^{22}Na in channel 1. This calibration constant gives the c.p.m. per unit activity of ^{22}Na.
$C_{12} - C_{22}$ = similar definitions as C_{11}.

To calculate activities of ^{22}Na and ^{42}K from the equations above one must first determine the calibration constants. This is done simply by measuring known activities of ^{22}Na and ^{42}K separately. By dividing the c.p.m. by the known activities the c.p.m. per unit activity, which are the calibration constants, are obtained. As the calibration constants are

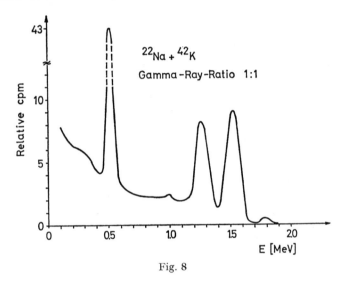

Fig. 8

the base for calculation of the activities they have to be known very accurately. The statistical error in counting the standards should therefore be kept small. This can be achieved by using not too small activities and adequate counting times.

With a measuring geometry of about 40% and using a 8 × 4″ crystal, the calibration constants for ^{22}Na and ^{42}K have the following values:

$$C_{11} = 60 \text{ c.p.m./nCi} \qquad C_{12} = 2.26 \text{ c.p.m./nCi}$$
$$C_{21} = 1.58 \text{ c.p.m./nCi} \qquad C_{22} = 8.35 \text{ c.p.m./nCi}.$$

The calibration constant which reflects the counting rate at the peak of the ^{22}Na spectrum is much greater than C_{22} which corresponds to the ^{42}K peak. This is caused by the fact that ^{22}Na yields a gamma quantum with each disintegration. With ^{42}K we have a gamma quantum in only 18% of the disintegrations. One therefore reaches about the same peak

heights if the activities of the two radionuclides have a ratio of 1:0.18.
A spectrum of such a sample is shown in Fig. 8. One can see the two peaks
from ²²Na and the one from ⁴²K. An activity ratio like this, which gives
about the same number of gamma rays from each radionuclide, is ideal
for such a measurement. Under these conditions, the errors in determining the two activities are at minimum.

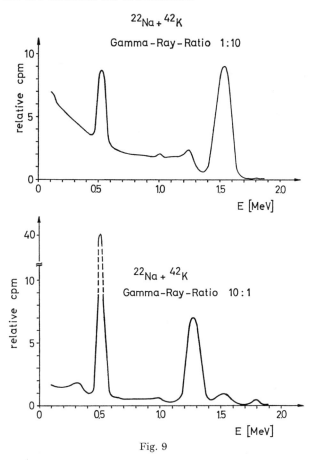

Fig. 9

A determination can however still be made if the ratio of the gamma
rays is much less favourable. This can be seen in Fig. 9. The upper
spectrum is from a sample where the gamma-rays from ⁴²K and ²²Na
have an activity ratio 10:1. This spectrum is thus mainly a ⁴²K spectrum
but the peaks from ²²Na can still be recognized. In the lower spectrum
we have a ratio of 10:1 in favour of ²²Na. But here also the peak from
⁴²K can still be seen.

An important question is with what accuracy the two radionuclides can be determined if the ratio of their gamma rays is not too favourable. The accuracy depends first of all on the total number of counts which have been counted in each channel. As the radioactive decay is a statistical process, the standard deviation is equal to $\sqrt{n}$, if n is the total number of counts. If we have a sample with 0.1 μCi ^{42}K and 0.008 μCi ^{22}Na (ratio of gamma rays 10:1) we have to expect the following count rates:

Channel 1: 226 c.p.m. from ^{42}K   Channel 2: 835 c.p.m. from ^{42}K
108 c.p.m. from ^{22}Na   3 c.p.m. from ^{22}Na
_____   _____
334 total   838 total

If the measuring time is 3 min, then in channel 1 about 1000 counts will be registered; that means the statistical error is 30 counts or 3%. In channel 2 we shall have about 2500 counts; here the standard deviation will be 50 counts or 2%.

The counting rates with their standard deviations are therefore:

Channel 1: 334 ± 9 c.p.m.
Channel 2: 838 ± 16 c.p.m.

To get an estimation of the possible error in determining the two radionuclides, one can make the calculation according to the Eq. (1) with the extreme values. That means with 343 respectively with 325 c.p.m. and with 854 respectively 822 c.p.m. ^{42}K can be measured with a standard deviation of 2% and the standard deviation for the activity of ^{22}Na is 6%. These are about the same standard deviations one would reach if the two radionuclides had been counted in separate samples in these two channels. This shows that even with an activity ratio 10:1 of the gamma rays an accurate determination of both radionuclides is possible.

It should be mentioned that in very favourable cases the calculation of the activities is even simpler then shown with the Eq. (1). This is the case if one of the two nuclides emits gamma rays which are much less energetic than a large fraction of the gamma rays of the second nuclide. Under this condition, C_{21} will be zero and then the activity of the second radionuclide can be determined without any influence of the first one. A good example for such a favourable case is the measurement of ^{22}Na and ^{24}Na. Fig. 10 shows the gamma ray spectra of the two radionuclides. If one works with the two channels between 0.3 MeV an 0.8 MeV and between 2.5 and 3 MeV, ^{22}Na will not contribute to the counting rate in the second channel. In such a case it would be good if the activity of ^{22}Na in the samples to be measured were much higher than that of ^{24}Na. Such a condition would give the highest accuracy.

In a completely analoguous procedure, double tracers with two beta emitters are determined with liquid scintillation counters. The best known example is the measurement of ³H and ¹⁴C. Fig. 11 shows the pulse height distributions of these two radionuclides. One can see that it is possible to choose the two channels in such a way that ¹⁴C can be counted without any influence of ³H. The same is of course true for any beta emitter which has a maximum energy greater than 160 keV. Also if

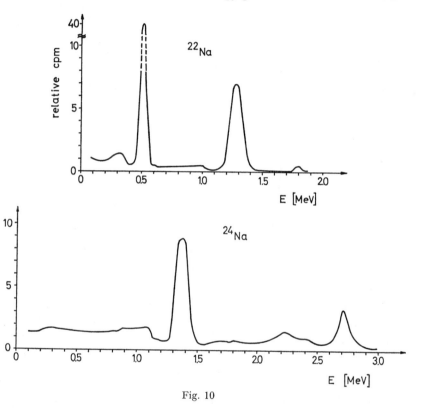

Fig. 10

¹⁴C and ³²P are in the same sample, ³²P can be counted without ¹⁴C interfering. Generally it can be said that if the maximum energies of the two beta emitters differ by more than a factor of 10, then the lower energy can be counted without any interferance by the higher energy. If the differences of the maximum energies are smaller the complete Eq. (1) has to be used.

Difficulties can arise if small amounts of quenching material are present in the sample and if these amounts vary from one sample to another. One cannot then work with the same calibration constants and they have to be corrected for each sample according to the quenching which occurs. To find the extent of quenching and right correction factor, three different methods are used in liquid scintillation counting:

1. Internal standard,
2. external standard,
3. channel ratio.

Only the first two methods are useful in double tracer techniques. 'Internal standard' means that after the measurement of the sample, known amounts of activities are added to the sample and then a second measurement gives the calibration constants. To make calculations easier, the added activities should be very large compared with the original ones. 'External standard' means that with an external source of ^{137}Cs a second measurement is done. According to the c.p.m. reached

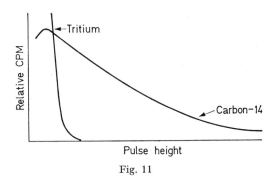

Fig. 11

with external ^{137}Cs a correction of the calibration constants can be made. But before this is done one has to know the relation between the counting rate of the external standard and the extent of quenching for the liquid scintillator to be used. Those measurements for the nuclides ^{3}H, ^{14}C, and ^{32}P are shown in Fig. 12.

In general, the measurement of beta emitters is more complicated than the measurement of gamma emitters. Since corrections for quenching are usually necessary, it is not easy to obtain the same accuracy as with gamma emitters. The point has to be taken into consideration if one has to measure in the same sample a beta emitter and a gamma emitter. Since the beta emitter very often gives beta particles also, a measurement in the liquid scintillation counter would be enough to measure both. But as it is easily possible to achieve higher accuracy by measuring the gamma rays with a Na I crystal it is often better to make a different measurement for the gamma emitter. If one does so, it is useful to arrange the experiment in such a way that the activity of the beta emitter is higher than that of the gamma emitter.

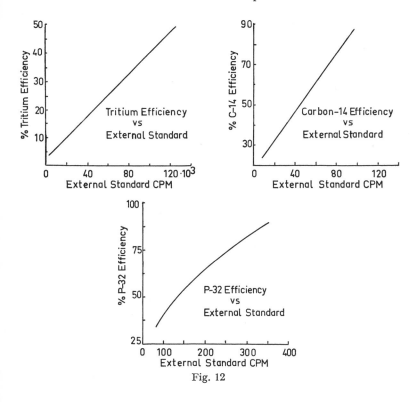

Fig. 12

References

BIRKS, J. B.: The theory and practice of scintillation counting. New York: Macmillan (Pergamon) 1964.

CROUTHAMEL, C.E. (Ed.): Applied gamma-ray spectroscopy. New York: Macmillan (Pergamon) 1960.

DAVIDSON, J. D., and P. FEIGELSON: Practical aspects of internal-sample liquid-scintillation counting. Int. J. appl. Radiat. 2, 1 (1957).

HEATH, R. L.: Scintillation spectrometry gamma-ray spectrum catalog, 2nd ed., Vols. I and II, IDO-16880. Phillips Petroleum Co., Atomic Energy Division, Idaho Falls, Idaho 1964.

HINE, G. J., and R. C. McCALL: Gamma-ray backscattering. Nucleonics 12, No. 4, 27 (1954).

OKITA, G. T., J. J. KABARA, F. RICHARDSON, and G. V. LeRoy: Assaying compounds containing ^{3}H and ^{14}C. Nucleonics 15, No. 6, 111 (1957).

PRICE, W. J.: Nuclear radiation detection, 2nd ed. New York: McGraw-Hill 1964.

RAPKIN, E.: Liquid scintillation counting 1957—1963. A review. Int. J. appl. Radiat. 15, 69 (1964).

WANG, C. H., and D. L. WILLIS: Radiotracer methodology in biological science. New Jersey: Prentice-Hall, Englewood Cliffs 1965.

Author Index

Page numbers in *italics* refer to the references

Druck: Carl Ritter & Co., Wiesbaden